OFFSHORE PLATFORM AUTOMATION

OFFSHORE PLATFORM AUTOMATION

By

M. A. Keyes

L. S. Rice

Process Controls and Optimization
Publishing Company
29801 Euclid Avenue
Wickliffe, OH 44092
(216) 585-8500

Editor: Larry Rice

ISBN: 0-9626691-0-5
Library of Congress Catalog Card No.: 90-061309

Typography & Printing by
D. Armstrong Co., Inc.
Houston, Texas

Dedication

We are dedicating this book to our parents . . .

Our Personal Parents

Marion A. Keyes, III
Winifred Nolte Keyes

Paul W. Rice
Dorothy J. Rice

Our Corporate Parents
Past
McDermott International

and Present
Finmeccanica Societa Finanziaria per Azioni

You have guided and given generously.
You have given materially, intellectually
and spiritually.
You have given sustenance and life itself.

TABLE OF CONTENTS

LIST OF FIGURES

PREFACE

The intent of this book is to provide general information about a new and relatively undocumented field—Offshore Platform Automation.

It is expected that the audience will extend from the investment community to operations management. Due to the breadth of this field and its anticipated audience, it is impossible for us to cover the entire realm of technology, applications, or areas of potential interest.

This field is relatively new and rapidly changing. While we have taken the present viewpoint wherever possible in the book, the progression of technology is impossible to predict, and where possible we have made some "best guess" projections.

It is anticipated that this book will become the first edition of several to follow. We hope to make a contribution to the industry by updating it as technology dictates. Should any errors or omissions be noted, please contact the editor: Larry Rice, Process Controls and Optimization Publishing Company, 29801 Euclid Blvd., Wickliffe, OH 44092.

While we have organized the book using a "macro-to-micro" approach, it is not anticipated that the reader will necessarily progress directly from beginning to end. It is suggested that the areas of specific interest be determined and investigated.

While many hundreds of suppliers, users, and consultants have provided invaluable information to the authors during the preparation of the text, we have taken care to remove most direct references to specific products, projects or organizations. In addition, no information of a proprietary nature has been divulged herein. All information used in this book is available in the public domain.

This publication is made available for information purposes only and shall not be considered a reference manual or instructional text. It is not intended for use with respect to engineering standards and practices, nor operating procedures and policies.

ACKNOWLEDGEMENTS

The preparation of any technical text requires the contributions of many people.

First and foremost, we would like to thank Don R. Warlick of Warlick & Associates for his continued leadership throughout this project. We would also like to thank Jeanette Gamse of Warlick & Associates for her efforts in the text preparation.

We would like to thank Linda A. Day, writer, for her tremendous writing ability and overall assistance in the basic preparation of the book. Without her contributions this book would never have come to fruition. We are also grateful for the editorial assistance of Mr. Edward J. Kompass whose contributions greatly assisted us.

Michael Clark's contributions in the artistic area are appreciated. We would also like to thank Linda Lister-Wilkinson for the cover design.

Many people within the Bailey Controls organization were instrumental in providing information and editorial assistance. Specifically, we would like to thank Don Dziubakowski, Claude Flandro, Bill Roach, John Blagrave, Doug Derry, Jim Cline, Dick Browngardt, Bill Napier, Phil Ryder, and Ted Gorrie for their contributions. We would also like to offer specific thanks to Jerry Stephens and his staff, Joe Patella, Ravi Kant, Don Frerichs, and Karan Vasudeva for their assistance.

Many of the users of automation equipment—customers, suppliers, and consultants spent many hours in interviews and discussions with the authors. Because they asked their names not be mentioned specifically, we will honor their request. However, our thanks goes to them all.

Last, but not least, we would like to thank Jane Rice for her patience and understanding during the long weeks of travel which were required for the worldwide research associated with this project.

—*The Authors*

Chapter 1

WHY AUTOMATE?

Successful businesses plan for the future before it arrives. In the past, the energy industry's greatest uncertainty was the whereabouts of new oil and gas; now the industry must also deal with the next OPEC pricing strategy and a seemingly unending stream of new governmental regulations.

With all the uncertainties, however, we know that three absolute certainties remain:

- In both exploration and production, cost containment coupled with increases in efficiency, operability, flexibility and productivity is the only way to survive pricing vagaries and increasing competitive pressures. The free-spending boom days of 1981 are gone forever.
- New reserves will be found in tougher places. As exploration pushes into severe ice regions and into deeper waters, economic development becomes possible only to the extent that we can eliminate the requirement for extensive human attention, and make our structures and systems invulnerable to capricious climatic conditions.
- Environmental protection and worker safety will continue to receive increasing attention, occasionally reaching what some consider paranoid proportions. New drilling, production, and transportation systems must reduce the possibility of human error to an absolute minimum.

Automation offers the greatest hope for dealing effectively with these challenges. This is particularly true in the offshore arena, where the environment poses almost as many operational difficulties as does outer space. In fact, the offshore industry owes much to the technological offshoots of space exploration.

This book examines the current state of the art of offshore automation and the management of the total production environment, working toward answers for such questions as:

- Why automate?
- What are the primary exploration and productions issues in various parts of the world, and how do they impact the decision to automate?
- What is the current state of the automation industry?
- What common concepts underlie all approaches to automation and total production management? In other words, what management processes are we trying to automate and how do they fit together?
- What are the components of modern digital distributed control systems, and how do they fit with the management processes we're trying to automate?
- What key elements of data acquisition, control, and diagnostic systems must be considered?
- What key elements of process control must be considered?
- What about facility management issues, including statistical process control, artificial intelligence, and expert systems?
- What applications lend themselves to automation?
- How do you implement an automation and production management system?
- How do you justify it economically?

We'll also include some case histories, and a look at some future possibilities for automation.

We'll start by looking closer at the question "Why automate?".

Cost containment

In the inherently risk-oriented energy industry, severe peaks and valleys are fundamental operating parameters. Success depends not only on efficiency in finding new oil and gas reserves, but also on the ability to keep costs as low as possible—or you're out on the streets the next time prices nosedive.

Today, the industry expects prices to take periodic nosedives, thanks to the immense role that OPEC plays in world markets.

By the end of 1988, OPEC accounted for 675 billion barrels of the world's total estimated recoverable oil reserves of 900 billion barrels—that's three-fourths of the world's reserves in 14 countries. In production, OPEC owns roughly half the world's capabilities. This might not be so daunting were it not for the fact that historical internal strife and squabbling among OPEC members makes predicting future oil prices largely an exercise in futility, a point which the market glut and price plunge of 1986 proved to everyone in the business.

Today, most people in the industry expect crude prices to range between $13 and $25 per barrel through the next decade and probably beyond the turn of the century. Major integrated oil companies and independent producers alike now make their business plans based on $15-per-barrel oil.

Today's leaders know the best way to plan for low prices is to utilize technology as efficiently as possible. Since prices started sliding in 1981, R&D budgets have increased as a percentage of gross revenue for producing, drilling, and service sectors alike.

Another factor that enhances the role of technology is the fact that oil-field development projects are long-term efforts involving an economic producing time-frame of 10 to 20 years. Fields that are marginal prior to a price decline must be retrofitted with the latest technology to remain profitable in the new environment.

Some major new technologies have appeared as a result of low prices and potentially still-lower prices in this long-term industry. For example:

- *Horizontal drilling* technology now makes it possible to drill sideways through an oil-bearing formation to increase production. Many fields currently producing and under development would not have been economically viable without this technology.

- *Multiphase flow*, the simultaneous transport of oil, gas, and water, may eliminate the need for much of the equipment now used on offshore production platforms to separate, meter, and treat these three components. The resulting simplification of offshore platform process equipment and structures could reduce costs dramatically.

- *Tension-leg platforms* (TLPs) using a bouyant platform restrained by anchored tethers create a stable and rigid structure that is vastly easier and cheaper to build and set up than conventional fixed structures, TLPs are most useful where water is deeper than 500 meters.

- Subsea completions, where one or more wellheads sit on the sea floor controlled and monitored from a central platform, reduce hardware installation and operating costs.

Advanced automation is another new technology whose time has come for the offshore industry. Simple remote monitoring and control systems have been in use for many years. Once considered sophisticated, today these one-step controls are only the prelude to a future that will inevitably incorporate the full power of artificial intelligence technology.

While driven by the same forces that propel the new technologies mentioned above, automation also has additional drivers:

- *Offshore people are expensive* to hire, pay and insure. Boats and helicopters to get them back and forth cost a lot more than a company car. Accommodations for lengthy offshore stays—including food service and recreational facilities—make the company cafeteria pale to insignificance. (For example, Unocal's unmanned platform in the Helder field of the North Sea cost 50% less to build and install than a conventional manned platform.) Safety considerations raise costs still further, and unplanned disasters, like hurricanes

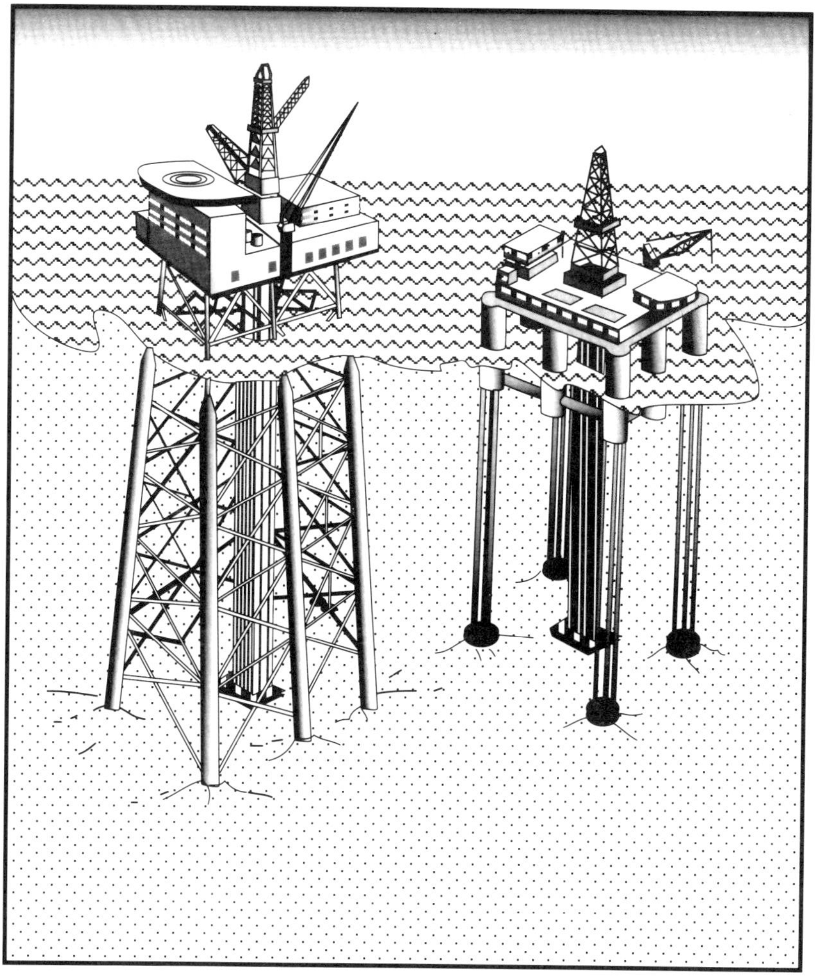

Figure 1-1. Conventional jacket platform and tension-leg platform

or labor unrest, can push costs out the roof while oil and gas languish in the ground.

- *People are not as accurate and efficient as machines* for many tasks they routinely perform. Machines are far more accurate and reliable in measuring and recording data, often allowing margins of error to be reduced from the 2.5% - 5% common in manual systems to less than 1%.

- *People are not as good in many maintenance functions as machines.* Computerized expert diagnostic systems can ensure that routine preventive maintenance takes place on schedule; they can monitor and often predict pending failures through advanced condition monitoring and Statistical Process Control techniques; and they can detect sudden failures and take immediate action to prevent such annoying irritations as ruptured tanks, explosions, blowouts, and fires. Moreover, many maintenance systems can be operated from a distance: Remote operators can determine the nature of malfunctions and prioritize maintenance activities, which saves money in transportation and labor. Currently, sophisticated remote monitoring and control systems operate to some degree in about 70 offshore installations.

So it's clear that new technology, particularly automation, plays an important role in keeping the lid on costs. And this role grows as the environment gets tougher.

Operating in hostile environments

While annual crude production climbed from just over 15 trillion barrels in 1969 to more than 20 trillion barrels in 1987, the percentage of crude coming from offshore fields rose even more dramatically, from 15% of the total to about 25%. Today, thousands of platforms dot the globe's continental shelves, producing close to 30% of the world's oil and almost 20% of the gas.

It's safe to say that most of the "easy" offshore fields have been discovered and developed. The Gulf of Mexico, a mild region with many reserves in relatively shallow water, now has more than 16,000 miles of pipeline coming to shore in 135 different locations. This contrasts sharply with sparse development in the severe environmental conditions of the northern North Sea. And even the North Sea pales in comparison to the iceberg-plagued environment of the potentially rich and almost completely unexplored waters offshore Alaska and Siberia.

Many factors can conspire to make a locale difficult for offshore development: wave height, currents, water depth, wind speed, temperature, storm occurrences, and sea-bed soil conditions. The cost of operating offshore increases in parallel with the severity of these parameters. That's not good news, because even in easy locations, offshore exploration and production is considerably more capital intensive than onshore:

- *Exploration rights* typically cost more to acquire offshore than onshore.
- *Geological data* is more expensive to gather, and relatively little archival data is available.
- *Expensive geohazard studies* must be conducted to understand the sea-floor characteristics in order to place a drilling rig and production facilities safety.
- *Deep-water drilling* rigs built for harsh environments are far more expensive: While day-rates for a deep well on-shore may reach about $10,000, offshore rig rates range from $12,000 to $70,000 and more (as of late 1989).
- *An expensive platform* is required to house production equipment.
- *Offshore personnel costs* and related expenses far exceed land-based costs, as discussed above.

Currently, offshore projects cost roughly 10 times more than onshore projects. This means that to be economically exploitable, offshore reserves must be much larger than they are on land, which, fortunately, they seem to be. The other factor that makes offshore developments economically possible is, of course, technology.

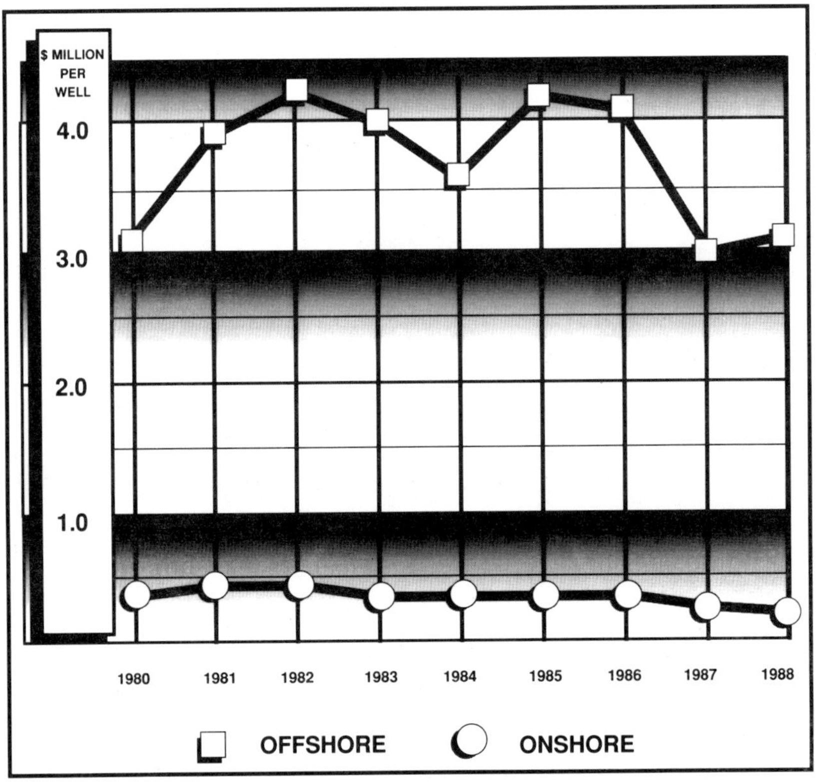

Figure 1-2. Typical drilling and completion costs, onshore and offshore

We are wading into deeper water in more difficult climates. And we're advancing only as fast as new technology can carry us.

Offshore technology started out slowly. The first offshore wells drilled south of Santa Barbara, California, in 1897 were placed on short piers extending from shore. In 1911, the first offshore well with no shore connection was drilled in Cadloa Lake, Louisiana. Barges supporting conventional land rigs made their debut in the 1930s as floating drilling rigs. In 1947, the first steel platform was installed in open waters of the Gulf of Mexico. And by the mid-1950s, about 100 rigs were at work in the shallow temperate waters of the Gulf of Mexico.

Early rigs in the Gulf worked at depths of about 100 feet, which grew to about 800 feet by the mid-1970s. Today's technology reaches still deeper to the ocean floor:

- *Shell's* 1987 Bullwinkle platform, a new version of conventional fixed-leg technology, extends 1,353 feet, taller than the world's tallest building.
- *Exxon's* Lena field guyed tower sits in 1,050-ft water offshore Louisiana. Like a radio broadcasting tower, it has the same width at its base as at deck level, and is supported by an extensive anchoring system.
- *Conoco's* 1989 Jolliet tension-leg wellhead platform (TLWP) is anchored in 1,754 feet of water in the Green Canyon field. The buoyant platform is tightly tethered to a sea-floor foundation template by a series of steel tendons.
- *Placid's* 1988 production platform floats in 1,500-ft water of the Green Canyon field. This production platform serves subsea wellheads in water as deep as 2,200 feet.

Today, waters must go beyond 2,000 feet to be considered "deep" for the drilling sector. The key is economic production from these depths. While it may be possible to produce from 5,000-ft waters with current technology, it is not economic. At present, the industry's drilling capabilities far exceed production capabilities in deep water.

The next step in going to deeper water is the further advancement of technology—not only in the design of new equipment to withstand severe environments, but also in the elimination of the human presence from that equipment. Since it is impossible (at least at present) to engineer a human being who will perform flawlessly for months or years at a time in sub-zero temperature and gale force winds, we must look for automated systems that will do the job.

Recognizing the importance of this task, the United Kingdom's Offshore Energy and Technology Board in 1988 outlined four major areas of research and development priority. Three of these areas involve extensive automation: development of an auto-

Figure 1-3. A typical offshore environment, a semi-submersible platform on the left, a conventional fixed type platform in the center, and a work boat/supply boat on the right.

mated drilling rig; advancements in subsea technology; and topsides weight reduction. (The fourth area is advanced systems for seismic data interpretation.)

Of course, automated applications exist already; for example:

- *Shell's* Bullwinkle conventional platform contains extensive automated process-control strategies to maximize production while keep maintenance and downtime costs to a minimum.
- *Conoco's* Jolliet TLWP requires complicated automated control of the buoyant platform's ballasting system. The integrity of this ballasting system is absolutely crucial to keeping the platform in place.
- *Placid's* floating production facility includes a highly complex electro-hydraulic control system to manipulate and monitor the subsea completion equipment. Reliability is critical for subsea completions because of the extremely high cost of manned inspection and intervention.

The value of these systems is indisputable: Automated systems in the Gulf of Mexico have often permitted operators to recoup the entire automation investment by avoiding only one hurricane production stoppage.

New systems under development carry automation even further. Perhaps the best example is Statoil's installation in the Veslefrikk Field in offshore Norwegian waters. The wellhead platform will be linked by bridge to a modified semi-submersible production platform, which will also be used for accomodations. A very sophisticated automation system will monitor and control the entire installation, including such application areas as drilling, production, HVAC, separation, treatment, and fire and safety.

While that offshore automation can help make today's typical "easy" fields more profitable, it becomes ever more important as we venture further out on the frontiers of deep water exploration and development. And when we consider environmental protection and worker safety, automation becomes mandatory.

Environmental protection and worker safety

The disastrous consequences of the events of 1988 and 89 have brought offshore safety to a new level of consciousness in the minds of leading offshore oil and gas companies. Upsets aboard production platforms have cost both property and lives. Perhaps the most famous disaster occurred in the explosion on Occidental's Piper Alpha production platform in offshore Scottish waters in July, 1988. While the investigation of that incident continues (as of late 1989), other upsets have occured, including an Arco platform in the Gulf of Mexico plus shutdowns on at least two Shell platforms in the U.K. North Sea.

Industry has responded to safety problems by inspecting more frequently, replacing dubious components, reducing headcount to reduce the number of lives at risk, and implementing more extensive automated safety systems. The move towards more advanced safety systems is so strong that these applications will inevitably be the vanguard of modern digital distributed automation efforts worldwide. We see this happening already in the North Sea.

Economic justification

Before you can automate any process, you must thoroughly understand its technical, political, and, most importantly, its economic dimensions: Can the project be justified within this framework? You must place a value on manpower reduction, product improvement, greater throughput, and safety enhancements. Understanding economic and other driving forces in the offshore environment is critical in deciding upon the automation design strategy most appropriate for your installation. In the offshore industry in the United States, the adoption of new technology has typically been done by operating management using traditional justification methods: simple calculations to demonstrate reduction in manpower or energy usage, reduction in product loss, or upgrade of product quality. These measures served well because they are simple to understand, relatively simple to use, and easy to explain to financial management staff.

For many years, these measures served quite well. During the first few decades of the offshore industry, technology developed at a relatively slow rate, and capital investments were easily justified because new technology had a relatively long life. Incremental improvements were seldom dramatic.

As the pace of innovation stepped up, more sophisticated approaches to project justification have evolved. The capital required to finance new developments continues to be justified by the old criteria, but new criteria now come into play, taking into account the realities of deep-water environments, severe climatic conditions, and over-riding safety requirements. These criteria are much harder to put numbers to.

The offshore industry began in the hospitable climate of the Gulf of Mexico in fairly shallow waters. In the late 60s and early 70s, many of the large oil companies viewed capital expenditures and project justification solely from a financial perspective. They used various standardized techniques to determine the exact value of specific investments; then they compared the worth of investment alternatives using net present value calculations or expected internal rate-of-return calculations. Recently, however, most sophisticated offshore operators are looking beyond these relatively simple financial approaches.

The time value of information

An added factor in justifying offshore projects has been the growing necessity to provide information in an ever-shortening time frame. The main factor driving this quickening pace appears to be the rapidity with which prices fluctuate for both raw materials and finished products.

The kind of information you need, and how quick you need it, make a big difference in the automation design strategy you adopt. A good automation design will provide accurate, real-time information not only to the local area operators for their process management and control decisions, it will also provide timely information for input to long-term operations, forecasting, and decision-support systems.

Some argue that today's information systems focus too intently on high-level management information at the expense of operating management. In the extreme case, this results in gathering of information suitable for statutory accounting needs, but not adequate for informed management at any level—much much less the dynamic needs of process control and operations. Conventional cost accounting practices that distribute overhead to direct labor hours based on a single platform-wide labor rate mask the true cost of individual products. Today, a good automation and real-time process-management system design includes the true cost of labor and benefits, raw materials, plus the price of finished products. Process management decisions thus incorporate all these factors, and this same useful information flows upward to senior management. This is extremely important on large integrated offshore platforms where several hundred people may be working in any number of disciplines on a variety of oil and gas projects. It is easy to see that the personnel mix, as well as raw material costs, can dramatically change the operational strategies of a platform at any given time.

A second problem that plagues the older information systems is the delay between the time a measurement is made and when it is actually available to the operator for decision-making. Since management decisions are often directed toward correcting the next measurement, the time delay reduces the time available to carry out that control function, which can result in a larger deviations from the target. Information systems must gather data and provide management control reports in a much more timely fashion than they have in the past.

Of course, it's obvious that timely information is worthless if it's inaccurate. Today it appears that the best way to gain both timely and accurate data is through new "smart" and "wise" sensing devices. These devices convert raw measurement data to the proper format within the time frame required by the measurement. The control system gets the information as quickly as possible for the required process control changes. And management gets the information they need to make timely executive decisions.

The time value of information has very strong implications in the cost justification of automation design strategies. Choosing when and how much automation to incorporate on an offshore platform is directly related to the cost of the investment and the required need for the information. Only a few years ago, it was difficult to justify a retrofit installation of process control based solely on the quality and amount of information that it could provide. But with the fluctuating prices of crude oil and the growing complexity of today's offshore platforms, the need for timely information has grown dramatically. This, coupled with recent tragedies on offshore platforms, provides a strong incentive for operators to use the latest technologies available.

In the past, retrofitting any aspect of an offshore platform was seldom done. Today, operators view projects to retrofit existing

Figure 1-4. A large land-based fully integrated process control system.

platforms with new integrated process-management, control, and safety systems as imperative to protect their personnel and investment.

Summary

The answer to the question "Why automate?" is that automation and total integrated production managment offers the only way that the offshore industry can advance into the future on a secure economic footing; the only way that reserves located in hostile seas can be reasonably exploited; and the only way to meet increasing requirements for environmental protection and worker safety.

The real question is not "Why automate?" but "How?"

Chapter 2

WORLD OFFSHORE ENVIRONMENT

Fraught with expense and risk, offshore exploration and drilling probably has the highest risk-to-reward ratio of any legitimate business undertaking in the world. Yet it remains a dynamic, vital business, in spite of recent turmoil.

The offshore industry got its real start in the temperate climate and shallow waters of the Gulf of Mexico more than 40 years ago. For many years, the Gulf was the only place where anything happened. But as oil was discovered elsewhere, the industry committed major financing in many other parts of the world.

In 1990, the North Sea is the leading producing area for offshore oil and gas. Latin America follows in second place, with the Gulf of Mexico not too far behind. The Middle East, Far East/Oceania and other smaller regions show some activity, and may become more dominant influences as today's most active fields play out. Deep-water activity is concentrated in offshore Brazil and several areas in both the North Sea and the Gulf of Mexico.

In this chapter, we'll review some of the factors that affect offshore exploration and production in the major geographic areas where it takes place.

Gulf of Mexico

The Gulf of Mexico is the world's most dense offshore drilling and production province, with a long history of success.

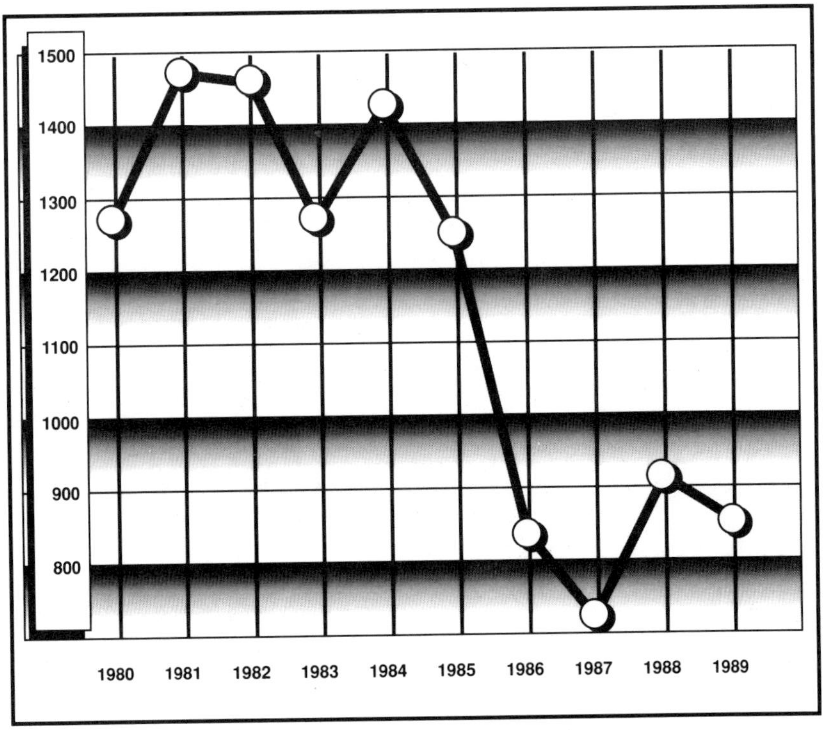

Figure 2-1. Drilling activity in the Gulf of Mexico

Drilling activity has remained relatively steady compared to other offshore regions, although it has followed world crude oil prices and overall offshore economics. Drilling peaked in 1981, when 257 offshore rigs were working in the U.S. and fell to less than 100 in 1987. By late 1989, the offshore rig count increased to 158 rigs working.

Today, there are more production structures in the Gulf of Mexico than in any other region—approximately 3,500 units. And the pipeline network supporting them reaches from eastern offshore Louisiana to Corpus Christi in South Texas, right at the Mexican border.

Salient features of the Gulf environment include:

- *Old and mature technology.* Most platforms here were built 25 or 30 years ago (intended for a 25-year lifetime), and they are old. The Gulf of Mexico is burdened with a large, aged infrastructure of offshore platforms and pipelines that were engineered, designed, and constructed in another, and much less rigorous and demanding technological era.
- *Shallow water.* Although new plays in the Gulf are developing in deeper water, the history of the area is based on shallow-water exploration in seas with low wave heights and predictable seabed soil conditions, with all operations relatively close to shore. The inherent "easiness" of this environment, coupled with ready accdss to an abundance of service companies and oilfield equipment manufacturers made exotic technology largely unnecessary in the past.
- *A (mostly) temperate climate.* Normally, the worst problems of the Gulf environment result from corrosive conditions, high heat and humidity. Occasionally, however, hurricanes devastate the area, so all facilities must be designed to withstand high winds and heavy seas. Historically, the response to a hurricane was to shut down production and evacuate the platform; automation now offers significant benefits by allowing production to continue, even when platforms are unmanned.
- *A low-cost mentality.* Because Gulf drilling and production goes back four decades, and because so many service companies compete for business in that area, there has been an emphasis for many years on doing business for the lowest possible cost. That includes engineering, design, installation, and platform operation.

  In the last few years, the Gulf has also seen increasing participation by smaller independent oil and gas companies. Many of these ventures buy old equipment removed from expired reservoirs and refurbish it for reinstallation in another location.

As a result of all these conditions, offshore operating companies in the Gulf have generally been slow to automate in the past. They have often used simple pneumatics or locally mounted manual controls. To achieve even minimal control and safety, operators had to be vigilant and thorough, check platforms and equipment frequently and use offshore personnel to monitor the performance of the few automated systems they operated. Even with the utmost in discipline and attention, these old manual methods could not assure safety or environmental security.

Fortunately, the old strategy of using imprecise, risk-prone manual methods is gradually giving way to automation. There are several reasons:

- The overall costs and liabilities involved with maintaining a sizeable offshore workforce dictates the need for more efficient and effective use of personnel.
- The possibility of producing from offshore platforms during storm or hurricane conditions is a reality today, and it offers significant economic potential.
- The complexity of new structures going into deeper water farther from shore requires some form of microprocessor-based automation.

We are already seeing early stages of a new implementation strategy that includes automation. By employing a fairly simple automated system on one of their production platforms, Arco Oil & Gas reportedly recouped their investment cost with production revenues generated during a 1988 hurricane. Conoco's Jolliet tension-leg wellhead platform (TLWP), which began production in November of 1989, has two separate automated systems on board: One monitors and controls buoyancy and tensioning of the steel tendons which tie the structure to the sea floor, while the other monitors and controls wellheads.

The current focus of automation efforts in the Gulf of Mexico are new automation systems for platforms going into deeper water and even consideration of some retrofit applications for existing platforms.

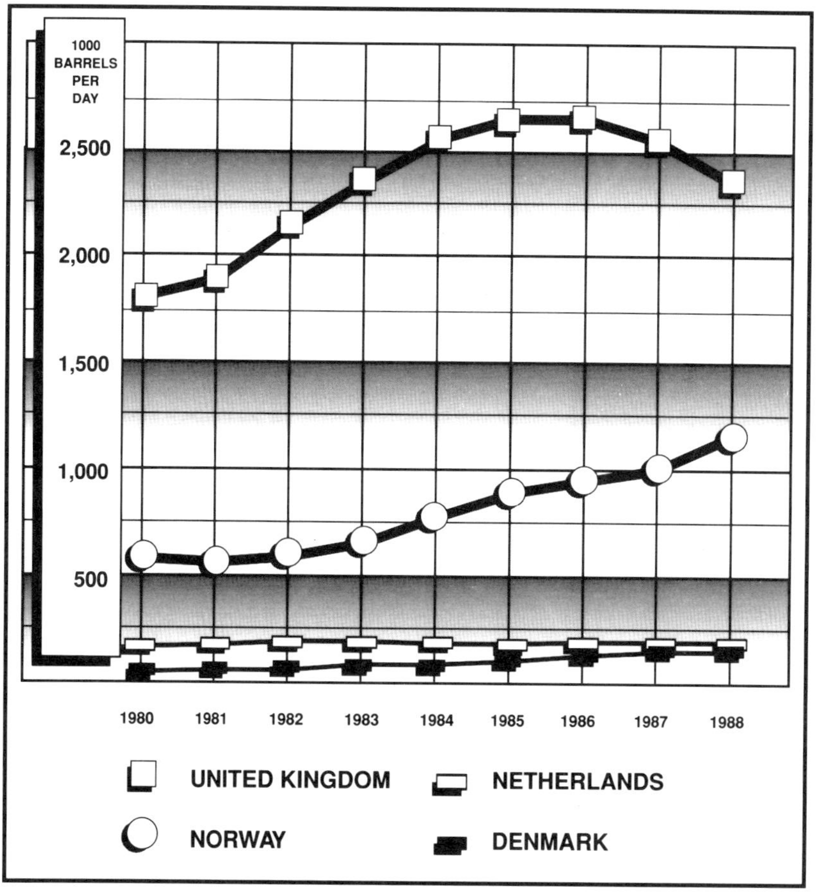

Figure 2-2. Production by North Sea producers

North Sea

Five countries drill and produce in the North Sea. In order by production volume, they are the United Kingdom, Norway, Netherlands, Denmark and West Germany. Each country has control over drilling and production rights for its own offshore area. In the United Kingdom, the U.K. Department of Energy distributes leases for tracts offered in licensing rounds, then allocates exploration rights. The governments of Norway, Netherlands, Denmark and West Germany operate in somewhat similar fashion.

Operations in the North Sea have developed rapidly over a very short period of time. The first licensing round was held in 1964 in the U.K. By 1971, total North Sea crude production reached 7,800 barrels per day. By 1988, it exceeded 3.5 million barrels per day. In addition to large oil and gas fields with high rates of production, other features of the North Sea environment include:

- *Extreme weather.* Low temperatures, mammoth wave heights and extreme storm conditions mean that companies operating here must be well-financed, committed to engineering and operating excellence in every dimension, and prepared to support long-term operations.
- *High costs.* The Brent Field development in the U.K. sector has four platforms plus separate oil and gas pipelines, and it cost approximately $7 billion to engineer and install. Drilling expenses in the North Sea can top $100,000 per day when fully burdened with all associated services and corporate overhead.
- *Massive, complex platforms.* The size and extent of the giant oil fields in the North Sea require a similar scope in drilling and producing structures. While early platforms resembled those in the Gulf of Mexico in size and design, most being installed today are larger and vastly more complex. In Norway, both the Gullfaks and Veslefrikk fields have major producing structures incorporating considerable advanced technology.

All of these factors make the North Sea an area where automation offers significant benefits. Add to this the emphasis on safety (especially since the Piper Alpha disaster), and you may expect the area to continue to lead in developing and implementing new technology.

Other offshore areas

(1000 Barrels per Day)

Region	Total
Latin America	3,232
North Sea/Western Europe	3,171
Middle East	2,651
Far East/Oceania	1,557
Africa	1,445
North America	1,257
Eastern Bloc	165
Total	13,478

Basis: Daily Averages 1986

The other offshore drilling and production provinces of consequence are Latin America, the Middle East and the Far East/ Oceania. In general, operating conditions in all these areas are quite similar to those in the Gulf of Mexico. However, since these areas are newer, it appears they may be good candidates for advanced technology, particularly automation, as new production platforms are installed.

Fields in the Far East provide one additional incentive for automation: produced crude tends to be of low quality, and applications that can ensure good wellhead separation and treatment offer significant financial benefits. This appears to be an application "made to order" for rapid subsurface completions and floating production systems, as well as on-line statistical process control.

We should see more automation in these markets because many of the most sophisticated operating companies who drill and produce in the Gulf of Mexico and North Sea are also present here and have technology to transfer.

In addition, state oil and gas companies with significant offshore operations are attuned to automation technology and receptive to its installation on future offshore structures. They are acutely sensitive not only to economics but also to the critical role that advanced automation and production management systems play in assuring safety and environmental security.

Summary

Obviously some offshore areas will move more quickly into fully automated installations than others, depending on the factors that support installation—the economics, the need for personnel reductions and increased staff effectiveness, and the complexity of the installation, for examples. Today, while operators in the Gulf of Mexico consider retrofits, others in deeper waters there, in the North Sea, and around the world demand the latest in automated systems to manage production, manage environmental risk, improve the quality of the product, reduce headcount, and bolster the effectiveness and safety of those who remain on the platform.

Chapter 3

THE OFFSHORE INDUSTRY TODAY

Within the past few years, automation has found its way into many processes that traditionally had been manually controlled. Panelboards, dials and gauges have given way to an onslaught of electronic instrumentation and computer and microprocessor-based terminals. Automation and expert systems applications have found their way into nearly all offshore areas: exploration, platform design and construction, drilling, production, subsea completions, support, pipeline, and storage and distribution systems.

Exploration

Seismic data acquisition and analysis remain the basic tools of geophysical exploration. As computers have advanced in speed and capability, so too has seismic data processing. Batch-processing, once the only method of processing, now often takes second place to microprocessor-based seismic workstations with on-line color graphics, three-dimensional subsurface analysis, and complex analytical algorithms like interactive refraction statics that increase the accuracy and sharpen the image of the earth's hidden geological structures.

Perhaps the largest role that automated monitoring and control plays in offshore exploration is in the area of mapping the "shoot," the gathering of the data. Ships gather data by trailing a string of hydrophones behind them as they generate the sound waves that ultimately create the seismic image. Accurate data requires constant monitoring and control of the ship's position,

Figure 3-1. A large multi-level integrated production facility.

generally through satellite-based automated locating and positioning applications.

Platform construction

Designing systems to bring hydrocarbons ashore stretches the imagination, intelligence, and practical abilities of everyone involved. As wells go into water over 500 meters deep in more hostile environments, the challenge increases.

Designs for deep-water structures depend both upon envimental conditions and construction-related issues. Environmental conditions include such things as expected storm conditions (wave height and period, winds, and tides), water temperature and chemistry, ice loads and frequencies, currents, bottom conditions, earthquake potential, expected useful life, projected deck loading, and, of course, the purpose of the platform and the extent of manning desired or required.

Construction-related issues include fabrication restrictions, location, material sources and quality, manpower quality, and the extent to which automated design techniques are available.

Deep-water structures pose special problems because loads are greater and platforms are larger and more complex with a greater height-to-width ratio. It is in the design of these structures that automated applications have found the widest use.

Today, the complexity of design almost mandates the use of Computer Aided Design (CAD), Computer Aided Engineering (CAE), Computer Aided Manufacturing (CAM), and the deployment of advanced artificial intelligence and expert systems. Together, these tools provide the only effective means of integrating the efforts of everyone involved with such projects: naval architects, plus structural, civil, electrical, and process control engineers, fabricators, installers, computer technicians, and others.

Immense computer power is required for the billions of calculations that go into the design of a single deep-water structure, and to manage these projects during fabrication, installation, and hookup. The most advanced engineering companies have complex design systems that include computer-assisted design with full use of artificial intelligence technology. These design systems may also include project scheduling, material takeoffs, purchase scheduling, and cost containment algorithms.

Artificial intelligence and expert system technology offer enormous potential for offshore design and construction. At present, at least one company offers a design package that incorporates general rules and engineering standards in its data base. This rudimentary form of expert system helps ensure that all engineering work done on a project conforms to the rules and standards set down.

Drilling

In the last five years, drilling technology has advanced more than in the previous twenty. The new technologies that make this possible are materials, electronics, and information systems.

New materials technologies include synthetic diamonds, ceramics, elastomers, and new corrosion-and erosion-resistant alloys. With the new polycrystalline diamond compact drill bits, (PDC bits) diamonds may now be a driller's best friend. These bits drill faster than conventional rock bits, and they last three times as long or more.

The second rapidly-moving technology is electronics. As wells go deeper, temperatures and pressures increase, requiring tougher electronic components. Today we have a new generation of well instrumentation available, and the pace of development shows no signs of slowing down.

The third major technology is automation. Today, digital systems are taking over what used to be the province of instinct and gut feel. They are particularly important to a completely new way of drilling—steerable drilling systems.

Steerable drilling systems can drill at almost any angle up to and including 90° under the earth, and the angle can be changed at will. Instead of turning the drill bit with a rotary table on the rig floor, downhole motors in the bottom-hole assembly will handle the job. In 1988, more than 700 wells were completed at angles greater than 70 degrees, more than twice the number in 1987. By 1992, annual high-angle completions will top more than 2,000, and ultimately they could involve perhaps a third or more of the total drilling market.

The advantage of horizontal drilling is that where a vertical well might only provide 20 feet of exposure to a producing formation, a horizontal well can travel hundreds of feet right through the formation, vastly increasing the productivity of a single hole. Horizontal drilling can turn uneconomic or marginal plays into profitable ventures.

The tricky part of operating a steerable system is keeping track of where it is and where it's going. This requires a measurement-while-drilling or MWD system. MWD devices go down the hole, conduct real-time directional surveys, and transmit the data back to a digital system on the surface for computation of drilling trajectory corrections. Here is an obvious application for artificial intelligence; someday the downhole assembly may make autonomous "expert" decisions about where to go next!

Another area where automation should play an increasing role is the combination of MWD directional technologies with MWD technologies for well logging and reservoir analysis. Until recently these two functions were being handled by different companies; the advantage in developing one downhole tool that can carry out both applications is obvious.

Offshore drilling involves another use for automation: monitoring the drilling vessel, as well as the well. Most exploratory drilling is performed on a mobile type of floating or transportable rig, whether it is a bottom-supported unit or a fully-floating, dynamically-positioned vessel.

Rigs that rest on the ocean floor come in two forms:

Bottom-supported rigs include both fixed platform and jackup types.

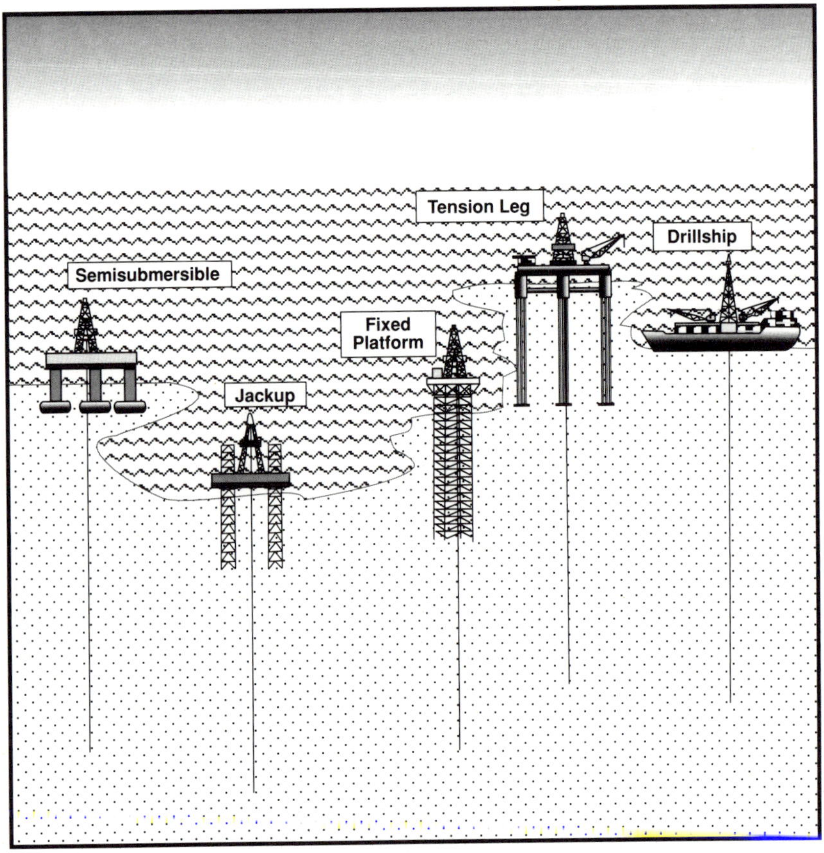

Figure 3-2. Various types of drilling rigs

- *Fixed platforms* are deployed by installing a large welded steel super-structure on the ocean floor. This structure, called a jacket, is a large pipe framework onto which are lifted the pre-fabricated modules.
- *Jackup rigs* can be utilized in deeper water and may be transported over long distances, either under their own power or on tow. The three or more legs of a jackup rig pass through the hull. Raised in the air when the rig is underway, they are jacked down to the ocean floor at the drilling

location, until the hull is completely and safely out of the water.

The deepest water requires a vessel-type drilling rig, either a drilling ship or a semi-submersible.

- *Drill ships* and ship-shaped barges are capable of drilling wells in very deep water where bottom-supported units cannot be used. However, since they float on the surface of the water, they are very susceptible to wave motion, making them unsuitable in heavy seas.
- *Semisubmersibles* are not as mobile, but are much better suited for drilling in heavy weather and waves. The semi-submersible drilling barge consists of a hull completely submerged in the water (where it is unaffected by waves) and an elevated platform above the waves supporting all of the drilling equipment. The semi-submersible can move under its own power or be towed to the position where the well is to be drilled.

Drilling vessels maintain their position over the ocean floor by spread mooring (using mooring lines and anchors), or dynamic positioning (thrusters in the bow and stern of the vessel that keep the vessel in position). More recent vessels use modern digital distributed control technology to combine these two techniques. Such a control system offers the additional advantage of full dynamic ballasting, so the vessel can hold position under extremely severe weather and wave conditions. (More about that later when we discuss applications.)

Control systems play a vital role in positioning. Sonar beacon systems use a submerged sonar beacon near the drill site, which acts as a homing device for the ship. Digital distributed systems or older computer-based position control systems monitor signals, add information about wave and wind forces, and control thrusters or mooring line tension to hold a precise position.

Recently, satellite positioning systems have been incorporated, allowing a ship to maintain positioning control so exact that deviations arise only when the combined power of thrusters and anchors is physically unable to resist wave, current, and wind forces to hold the specified position.

Production automation

In the early years (and on many older platforms today), producing from an offshore well was primarily a manual operation. This meant the regular monitoring and recording of crude gauge-indicated pressures, levels, and temperatures by the platform personnel. Later, pneumatic systems introduced a rudimentary level of remote operation and automation capability, followed by rudimentary electronic analog and digital logic control systems. These include single loop controllers and older programmable logic controllers laboriously programmed in the relay "ladder" logic originally developed for automotive manufacturing applications.

Now more than ever in the offshore world, safety shutdown systems, natural gas compressors, fire pumps, gas turbines, generators, power distribution and protection systems, oil metering systems, and air compressors must be totally automated and managed in an integrated environment.

Offshore hydrocarbon production is, of course, very similar to on-shore production. The oil, gas, or mixed-phase stream comes from the well through a wellhead system that monitors pressure and flow and directs the stream to the processing equipment. Although each offshore platform almost always serves multiple wells within a close proximity (unlike most on-shore installations where wells are widely separated), the stream is conditioned and processed very much as it would be on-shore, and many automation techniques apply in both environments.

A typical moderately automated platform will have a control room with a bank of process control equipment. In simpler installations, you might see racks of programmable logic controllers with their associated input/output devices, plus panelboards and perhaps a CRT with abstruse ladder logic data displayed. On a larger or more complex platform, you'll more

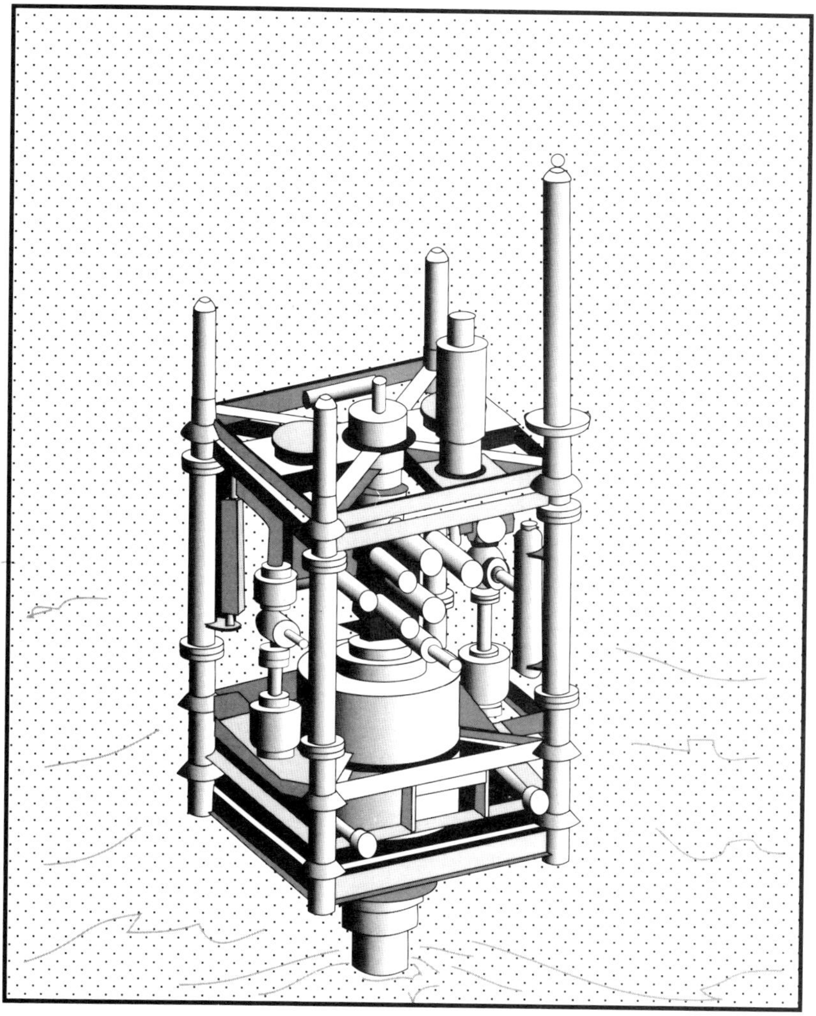

Figure 3-3. A typical sub-sea completion

likely find a full digital distributed control system with sophisticated CRT-based operator consoles.

Sub-sea completions

One of the most exciting new offshore developments is subsea completion. Subsea electro-hydraulic and hydraulic control systems have proven to be both reliable and cost-effective for distances of less than 100 miles. In the future, these distances will not be limited by hydraulic transmission of power, since dry, self-contained measurement and control modules will replace hydraulic lines.

An umbilical cord handling data transmission, plus hydraulic and electric power, stretches between the subsea template at the well(s) and the platform. At the template, there are separate hydro-electric control modules for each well's tree, plus other modules for the template and manifold valves.

Sub-sea completions are controlled by a master control station on a nearby platform. The master control station processes serial data relayed from a array of subsea sensors; interfaces with operators; and sends operational commands back to the sub-sea system. Sub-sea completions usually employ dual redundant systems that can transfer control automatically with no loss of data or control.

Support systems

On all platforms, a number of support systems keep things running and create a comfortable environment for employees. The most important support systems are:

- *Power generation:* Except for the main gas export compressors, which are driven by gas turbines, all services and operations on most of the newer platforms depend totally on electrical power. The generating capacity on some platforms is significant, sometimes in excess of 60 MW—enough to supply a medium-sized city. Should the system fail, emergency standby generators and battery backup systems on the platform must supply emergency electrical services immediately and automatically. All emergency and critical support systems must be fully backed-up.

- *Water:* The sea-water treatment unit provides a supply of chemically treated and filtered sea water for use as fire water, for process cooling, equipment cooling, and for injection into the reservoir. Potable water is supplied by distillation or reverse-osmosis systems that must also be monitored and controlled.

- *Diesel:* Crude fractional separation of produced hydrocarbons right on the platform can often produce sufficient diesel fuel (a light cycle oil grade cut) for power generation and standby generators.

- *Inert gas:* The inert gas delivery system is designed to purge vessels and pipelines of hydrocarbon gases and provide an inert "safety blanket" for tanks containing hazardous chemicals. The gas used is usually nitrogen.

- *Chemical injection:* The chemical storage and injection system is extremely important on a platform. A wide range of chemicals are used in both drilling and production systems: mud additives, spotting agents, biocides, anti-foaming agents, corrosion inhibitors, demulsifiers, etc.

- *Plant and instrument air:* The plant and instrument air handling system provides a constant supply of clean, dry compressed air for powering pneumatic actuators, for cooling certain instruments and controls, and as a transport medium for powdered materials. This air must be clean and dry because it is often used to purge enclosed analyzer compartments and other areas where potentially dangerous hydrocarbon vapors should not accumulate.

- *Heating, ventilation and air conditioning (HVAC):* The HVAC system provides controlled air throughout most of the platform. It also prevents accumulation of explosive gases in enclosed or confined spaces, which is extremely important for the safety of operators and the protection of the platform.

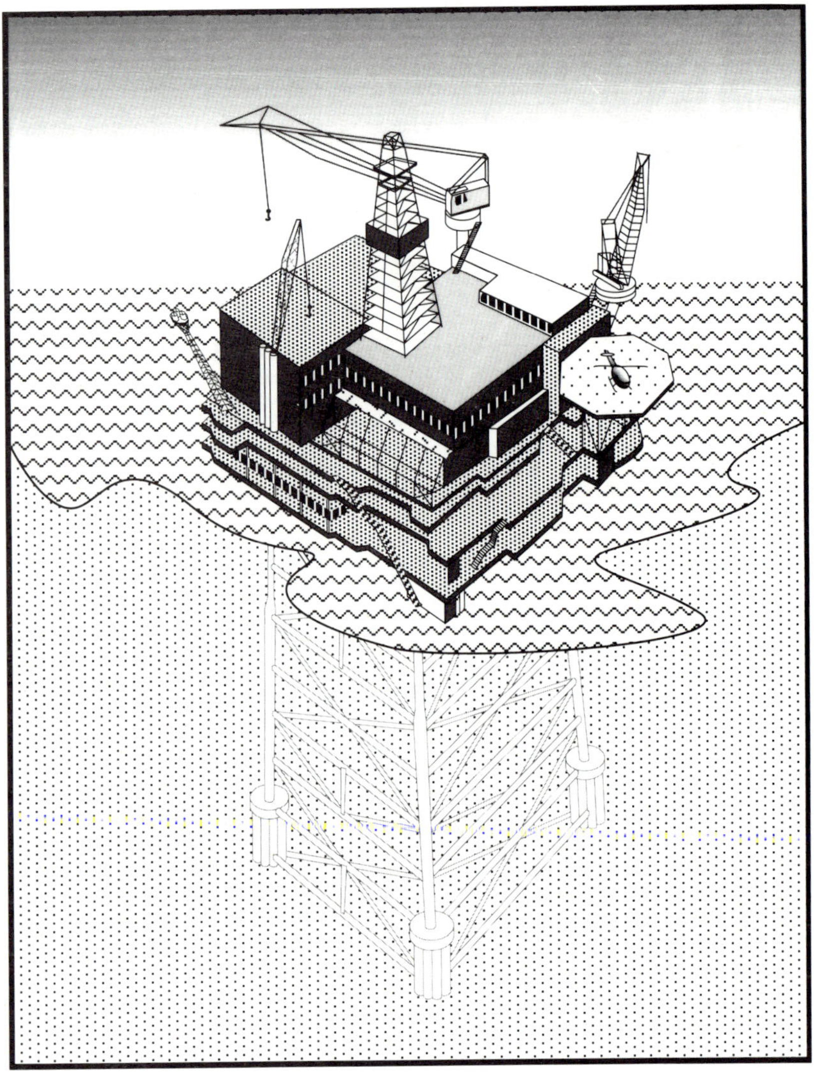

Figure 3-4. Typical large production platform showing supporting structure below the surface of the water. (Note: Compare with the fixed platform photos in Chapter 10.)

- *Hot water:* Hot water is generated either in a package boiler or in a heat exchanger on the exhaust gas of one of the gas turbines. Hot water is used by the HVAC system, for domestic uses, and for maintenance hot-water supply.
- *Helicopter Fueling:* The helicopter fuel storage system stores aviation fuel for the helicopters that provide transportation to the platform. Because supplies must come from an external source, inventory must be carefully monitored.
- *Drains and oily waste treatment:* Recent environmental protection regulations make monitoring, control and logging of wastewater treatment processes and effluents a critical concern of platform operators. The waste treatment facilities allow only oil-free liquids that fully meet EPA specifications to be discharged into the sea.
- *Sewage:* The sewage treatment facility is separate from process wastewater treatment, and is also geared to meeting environmental requirements for discharge of human wastes.
- *Flushing, well testing, and management of vents, flares, and drilling waste:* When a production well has been completed, it will contain a mixture of materials—oil, gas, water, sand, drilling mud, and other contaminants. If this is allowed to flow directly into the main process system, the processing equipment would quickly become fouled. The initial output must therefore be cleaned by flushing each well before it is brought on stream and disposing safely of all the waste products. Excess gasses must be flared safely.
- *Safety:* In today's social, political, business and operations climate, there is no greater concern than the safety of humans on the platform. All platforms have extensive safety systems and procedures.

  On some of the larger platforms, a standby vessel remains at or near the platform continuously. In the event of a serious upset, this boat can evacuate all personnel.

  In addition to evacuation ships, all platforms have sufficient lifeboats to rescue more people than can be housed in the accommodations areas and to ensure that contract employees and visitors can always be provided for in any emergency. These lifeboats are specially designed to withstand damage by fire or heat, and to cope with difficult conditions at sea.

New regulations in some areas require continuous on-line monitoring of the number and location of all personnel on the platform. Many of today's digital distributed control systems encompass this function.

Fire extinguishers are positioned at all key locations. Strategic enclosed areas are protected by halon gas systems; halon is a very effective extinguishing agent. Halon systems are controlled by the fire and gas detection system, and can be manually or automatically tripped.

Two systems monitor for hazardous events: The fire and gas detection system can provide early warning to activate a sprinkler or other extinguishant system, and the emergency shutdown (ESD) system can stop any part of a process automatically and in the prescribed manner.

Pipelines

Another sophisticated system which must be monitored and controlled are the pipeline systems. The pipline systems are often used to transport the crude oil or natural gas from the platforms into gathering networks and ultimately to shore. Up until very recently, these intallations were fairly unsophisticated and most of the monitoring was done for custody transfer only. However, as the number of pipelines have risen dramatically so has the need for process control and enhanced data acquisition.

In the case of natural gas, there are two vital sets of information which are required. The first is the physical flow of gas in the system at any given time. The second piece of information which is of great importance is the contractual allocation of gas at the inlet and outlet points of the system. Sophisticated systems which are available on the market today incorporate artificial intelligence, rule-based systems, which are used to allocate the volumes of natural gas within the pipeline system and determine how much of the total volume of gas should be assigned to any of the number of contracts supplied. While these accounting values have always been important as operators of pipelines have begun transporting gas owned by others, the need for sophisticated and accurate accounting systems has become exceedingly important.

Because crude oil is less volatile and generally under lower pressures than natural gas, these pipelines tend to be less sophisticated and less automated. Nonetheless, with the advent of great sophistication in the areas of process control and the implementation of embedded artificial intelligence and expert systems, the area of pipeline automation is ready for the dramatic improvements in process control and data acquisition which are available today.

Chapter 4

INTRODUCTION TO AUTOMATION CONCEPTS

Exactly what is automation?

A look at the forces prevailing in the marketplace convinces us that increasing automation, leading to total production management, is fast becoming a necessity: Automation offers a way to control costs and increase productivity in a world where the price of oil can fluctuate daily; it provides the tools we need to carry exploration and production into hostile environments; and it is the most effective means of improving safety and minimizing the risk of adverse environmental impact.

When we look more closely at the potential of offshore automation, we realize that the decision to automate is not a simple one. Needs differ in each part of the world: North Sea operators are concerned about safety in a dangerous climate. Gulf Coast operators worry about retrofitting aged platforms and evacuation during the occasional hurricane. In the Far East, the labor situation reduces the importance of demanning, while operators look for better wellhead production to make the difference between profit and loss.

Not only do regional considerations force different outlooks on automation, so does the current state of existing offshore platforms. As we have seen, extreme diversity prevails. The level of automation ranges from a complete absense thereof to extensive and sophisticated systems on some of the newer platforms.

Before we can deal intelligently with the automation issues that each application area raises, it will help if we have a solid idea of exactly what automation is.

Because the automation level of most production platforms is fairly elementary today, there is no good general understanding of what is meant by the term "automation." Some think of automation as simple remote data acquisition and control, while others see it as a system that manages a whole facility. Many think of pneumatic and electronic controllers as "the technology" of today, while others realize that the digital distributed control systems that have appeared the last few years offer far greater reliability, ease of use, safety and economy.

So what is automation? We take a broad view—that automation covers everything from the sensors to senior management. It includes data acquisition, monitoring, control, communication, statistical process control, artificial intelligence, expert systems, facilities management, and long-range planning. Put simply, *automation is the formalization of the management mission and process at all levels.*

This is no simple job. And the offshore environment makes it more difficult. Onshore, if a weld cracks in your plant, it is easy to evacuate personnel while you fix it. The risk of loss of life is generally much lower than it is offshore, where evacuation is a major event. In the offshore industry, we have ordinary processes in an extraordinary, confined, and unstable environment. There's not a lot of room for error or inattention.

Although we have said that offshore automation can encompass everything from the sensors to senior management, let's start with a simplified approach to an individual platform process.

A quick look at process control

Like any industrial facility, an offshore platform must perform many different processes to separate, transform, and transport raw materials and finished products. In each process, several actions take place:

The first is acquiring data from sensors. Like nerve endings in the human body, process sensors retrieve physical data such as temperature, pressure, flow, and level, and change it into electrical signals, with varying degrees of sophistication.

These electrical signals travel through a network to reach the logic of the system, in much the same way that our sensory data travels through the nervous system to reach the brain.

The brain of a process control system may reside in one computer system; this was the first means of bringing the benefits of digital process control to all fields of endeavor. More recently, control power is distributed among hundreds or even thousands of microprocessors throughout the plant. This approach gives greater reliability, control power, and safety at a lower cost.

Detailed system structure depends upon how complex the process is, how many processes are being integrated within a single process control system, and the physical layout of the processes to be controlled. As we'll see, with today's technology, you can carry integration just about as far as you want to. We'll call the brains the "process controller,' recognizing that the physical reality may be one or many microprocessors or even a computer.

The process controller

On the simplest level, the process controller takes signals coming in from the field sensors (e.g. a temperature) and uses preprogrammed algorithms (e.g. a limit check like "Is this temperature above the limit?") to make decisions about whether or not to act, and what to do. If the controller decides to act, it provides the appropriate signal to an output device. The output device is analogous to a muscle; it opens and closes valves, moves things mechanically, or turns something on or off electrically to influence the real world. In our example, if the temperature is too high, the system might turn down the steam or it might start a complex procedure to shut down the entire process.

Obviously, the system must all work as a unit. In the absence of redundancy, if any single part does not operate correctly, the

Figure 4-1. A stand alone process control system

system as a whole will not function. (This observation points out the critical need for redundancy in offshore applications.)

One of the most important aspects of an efficiently functioning system is the communication network. In the world of process control, the value of information depends heavily on its timeliness. Often, the more timely a decision, the more cost-effective, safe and secure the result will be.

The communication systems developed for offshore platforms over the years have taken several forms: pneumatic, analog, and digital.

Pneumatic communication systems

The early systems were pneumatic—tube bundles filled with pressurized air. Changes in pressure could reflect either changing values detected by sensors or changing signals sent to actuators.

Pneumatic systems do not have a high degree of accuracy because exact digital values must be converted to an analog air pressure in a fairly narrow range between 3 and 15 psi. Undetected air leaks or dirty air can also affect accuracy. On the plus side, pneumatic systems are quite rugged (though not as reliable as current digital electronic systems) and do not require electricity at sensors or actuators. (Improperly designed electrical or electronic systems could pose a potential hazard on an offshore platform.) Pneumatic systems are still widely used.

Analog communication systems

The initial electronic systems were analog, a signal from 4 to 20 milliamps or 0 to 10 volts transmitted over a twisted pair of wires or digital electronic replacements for older relay logic control systems. (The meaning of this is unclear, especially because we haven't talked about relay logic yet.) Electronic systems evolved in two directions: analog and digital.

Analog systems also took two roads: single loop analog controllers and programmable logic controllers (PLCs):

Figure 4-2. A pneumatic system

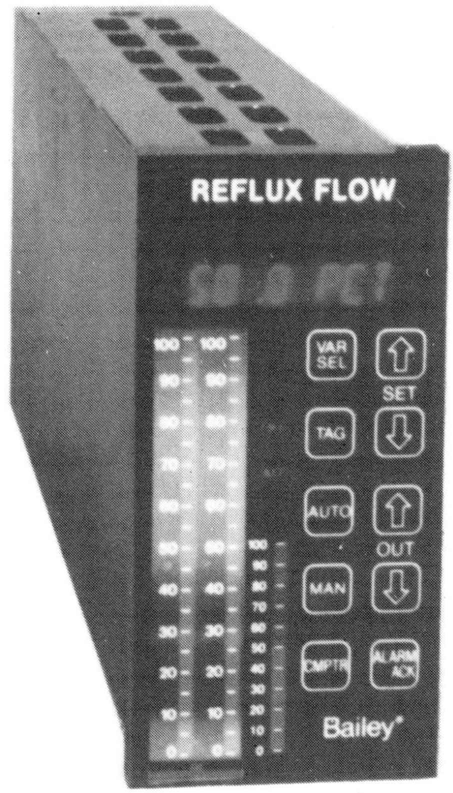

Figure 4-3. Typical single-loop controller

- *Single loop controllers* were developed by process and control engineers and designed to use basic proportional, integral, and derivative control strategies to regulate simple process or parts of process.

- *Programmable logic controllers* were developed for the automotive industry, with electrical engineers and maintenance personnel in mind. PLCs use ladder logic to operate relays and timers. In the past, PLCs have been limited to simple on/off capabilities, but more recent advances have expanded their regulatory capabilities, although they are still rudimentary compared to full-function digital distributed control systems.

Today, all types of these electronic systems are still in wide use.

Digital communication systems

Digital control technologies began to gain attention in the 1970s. Distributed control systems based on digital microprocessors opened the door for digital communication protocols of significantly higher speed than those previously used. This architecture provided distinct advantages, because it allowed quicker analysis of incoming data and quicker response with outgoing data. It also provided the perfect opportunity to analyze a lot more data and control more aspects of the process in a far more efficient, effective and safe manner.

Microprocessor-based systems have revolutionized process control by putting computing power previously reserved for large centralized mainframes right out on the plant floor. In some cases, considerable computational capability has been built right into sensors themselves, in the form of "smart" or even "wise" transmitters.

Now let's look at some of the specific factors driving the current technology explosion in offshore automation.

The evolution of new automation concepts

Digital electronic and control technology has advanced rapidly in many industries, producing a new generation of instrumentation, communications, and control systems that are now finding their way offshore. Some of the concepts driving this technology evolution are described below.

Integration of systems

Today's digital technology allows a single consistent system to perform nearly all monitoring, safety, and control functions.

And it's a good thing. Having fewer windows to all platform operations reduces operator confusion and errors, simplifies maintenance, lowers the cost of the spare-parts inventory, and prevents potential hazards (such as the fire-fighting system trying to put out a fire while the HVAC system is blowing fresh air into it). Ideally, one should have a single window to all systems on the platform. That level of integration is consistent with the capabilities of the most advanced of today's systems and is now being deployed by progressive offshore owners and operators.

The applications that can be integrated and/or handled jointly on a platform fall into four major subsystems:

- Process control and accounting systems for production, treating, and custody transfer of oil and gas;
- Structural, environmental, and meteorological monitoring;
- Utilities control (Power, energy management, HVAC, water); and
- Safety and emergency shut-down systems.

An important part of integration is the replacement of panelboards with CRT technology, which offers greater security and freedom from potential operator error, greater flexibility, faster response and more information, all with less weight and space. Standard CRT formats also reduce the different types of devices, data formats, and procedures that operators have to know and use, simplifying both training and operation.

Integration also has the benefit of reducing maintenance demands, spare parts requirements, and cost.

Distribution of intelligence

While artificial intelligence was once the sole province of the ivory tower, today it is spreading contagiously to an increasing array of industrial processes. "Expert systems" are the most common current application area for artificial intelligence technology in process control systems.

Figure 4-4. The control room of a large integrated facility.

Expert systems typically use a mix of conventional analytical methods, algorithms, and knowledge bases specific to the application, plus rules and data that may even be imprecise or "fuzzy." With these resources, a powerful "inference engine" mimics intelligent behavior and draws conclusions or takes actions appropriate to the current or forecast environment.

While it is stretching the truth to classify these systems as intelligent in the human sense, they are nevertheless useful, especially in such mission-critical offshore applications as:

- "Smart" diagnostic, failure prediction and emergency-action systems.
- Control systems that require pattern recognition, or applications that are intractable to normal analytical methods, such as integrated equipment start-up, monitoring and shutdown.
- Access monitoring, control, and security systems.
- Inferential three-phase flow monitoring and control.

- Systems where possibly-conflicting data from more than one source must be evaluated, integrated, and used to provide optimum control.
- Highly optimized, integrated production and maintenance planning systems.

These systems have now been successfully integrated into today's most advanced digital distributed systems, removing many of the key objections and obstacles to process-control applications of artificial intelligence technology. The new integrated control/expert systems offer many advantages over earlier attempts to "attach" artificial intelligence to existing process-control systems:

- Integrated systems do not require separate, typically unreliable, artificial-intelligence hardware and software for their use.
- They don't require human experts on artificial intelligence to tend to them during operation.
- They don't depend on slow serial communications between an external processor and the process control system.

Control system applications today are no longer limited to what happens the control unit. They are spreading rapidly outward to "smart" and "wise" transmitters, positioners, and valves.

The case of transmitters is particularly interesting. Originally transmitters simply converted temperature, pressure, or flow into an electronic or pneumatic signals linearly related to the engineering value of the process variable. Early enhancements provided signal filtering and rudimentary transformations, such as taking the square root to determine flow from differential pressure across an orifice.)

"Smart" transmitters can adjust the span and zero remotely, identify each transmitter by its tag number, and provide more complex transformations. And today's "wise" units have an even greater infusion of imbedded artificial intelligence technology at their command.

"Wise" transmitters can not only infer the presence and cause of failures, they can report these failures and take appropriate action. They can also simplify routine tasks such as setting the output level to a value that depends on the application, the engineer's preference, or the circumstances. The value of that output level may come from a number of sources—the last reasonable value; a predetermined value; high, low, or full-scale values; etc.

"Wise" transmitters also have other enhanced features. Logical filtering capabilities make them more efficient in separating signals from noise. And enhanced communication capabilities allow them to transmit process variables in either conventional analog or digital, dedicated or bused architectures.

No discussion of artificial intelligence as it relates to process control is complete without a brief description of the structure and the roles of data, information, knowledge, and wisdom in such systems.

- *Data* is the raw number—a pneumatic or electrical signal from the sensor. By itself, it is virtually meaningless.
- *Information* is data that has been checked for validity, scaled, linearized, converted to engineering units, and perhaps checked against upper and lower limits. It means more than data, but still doesn't tell us what to do.
- *Knowledge* is information that has been combined with operating and design experience, control algorithms, and technical objectives to yield some understanding about its relevance. Most expert systems (we'll discuss them in some detail later) work from a knowledge database.
- *Wisdom* is knowledge that has been combined with external constraints, a preset strategy, skilled assessment of inferred platform states, a forecast of future conditions, plus the overall strategy of the platform. Wisdom is the output of an expert system and can be the guiding force for both individual processes and the integration of multiple processes into a complete facilities automation system.

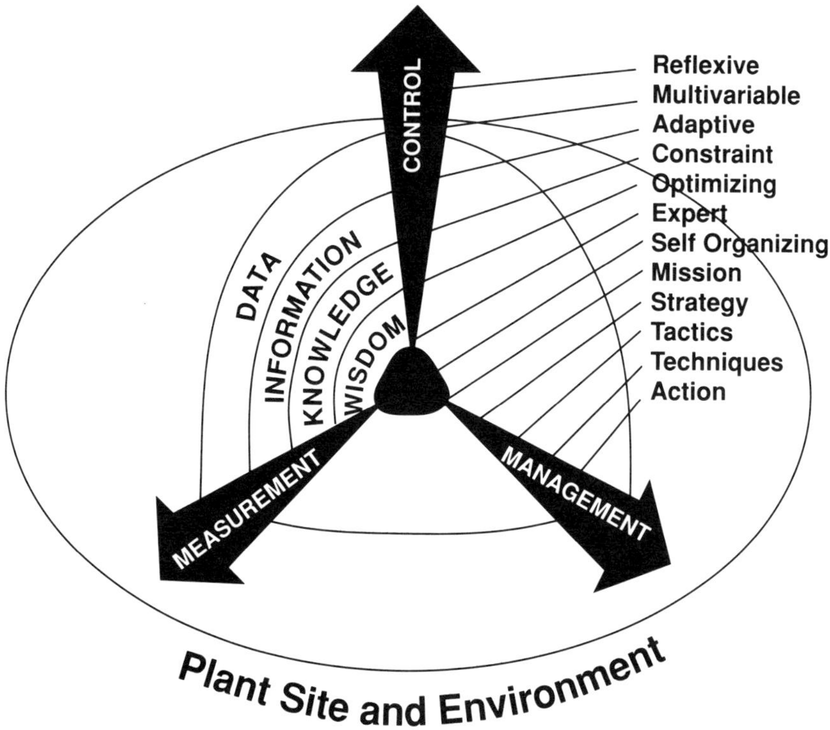

Figure 4-5. The relationship of data, information, knowledge, and wisdom

In this context, what makes expert systems so attractive in offshore automation is the ability of today's control systems to distribute tasks to the functional level where they are most appropriate. For example, the new generation of "wise" transmitters convert data into information right at the sensors. The further winnowing of knowledge from that information can start in the "wise" transmitter and conclude in the processing modules that serve each process and group of processes. And the final enhancement to wisdom can be accomplished by an embedded supervisory expert system. As we move up through a distributed system, each level of intelligence is appropriate to the particular tasks performed. The system has the ability to draw upon all types of resources at all levels; infer missing data, evaluate alternatives, and make recommendations or the take the appropriate action directly.

Reliability improvements

New automation systems exceed previous standards for performance, operability, reliability, and safety, while reducing cost, weight, and size. Reliability and safety, in particular, are critical to successful and economic offshore operations.

Reliability improvements as the plant or platform operator sees them can be found in two sources: within the process control system itself, and in new reliability-oriented applications for process control systems. Some of these applications include:

- *"Smart" or "wise" transmitters.* These data-acquiring devices monitor and diagnose their own performance by checking the quality of their outputs: looking for outputs that are unreasonably high or low or are changing unpredictably. Sophisticated versions can even check the frequency components of their output signals to see if they fall within normal limits. By incorporating auto-zero and auto-span setting functions, "smart" transmitters directly simplify maintenance and improve reliability.
- *Automatic diagnostics.* Located within the separate modules of a distributed system, built-in diagnostics are particularly useful in remote or inaccessible installations like subsea production facilities. Combined with self-repairing, self-organizing, or fault-tolerant system design approaches and expert systems, automatic diagnostics improve the quality of information and minimize lost production time. (Not to mention costly subsea inspection and maintenance time).
- *On-line, integrated structural analysis.* This function is becoming increasingly important with new platform designs. Acoustic monitoring is already in use to monitor flaw propagation and incipient failures in such process equipment as boilers, reactors, and turbines, and it can provide

early warning for problems in structural members. It can also generate the data needed to compile a stress history of the structure and forecast the remaining life and the requisite allowable maintenance horizon by analyzing wave forces, storms, ice loads, etc. Expert systems can be particularly useful here in suggesting courses of action to be taken, depending on the parameters monitored.

- *Equipment analysis.* Modern digital distributed systems use a number of techniques to produce an integrated equipment diagnostic system, including vibration and displacement monitoring, Fourier transforms, and power spectral density analysis. Combining predictive maintenance algorithms and condition-based maintenance with failure diagnostics in this way optimizes maintenance and reduces costs. We'll have much more to say about equipment analysis, both in the section on reliability considerations for distributed control systems, and in the condition monitoring section in the chapter on automation applications.

Reduction in engineering costs

The costs to design and engineer advanced automation systems for offshore applications can be reduced using new workstation-based engineering tools in conjunction with a design "language" that works like engineers think. The most advanced systems today are extremely easy to use, and are also self-documenting, so it's possible plot out revised drawings whenever you make revisions. In addition, once you have designed a process control strategy, you can down-load it directly to the process control system. You can also up-load it to check validity and verify "as installed" documentation.

Systems like this make it relatively simple to design complex automation strategies using standard, proven modules that can be combined in different ways to meet the needs of different platform operations. This technique can save thousands of hours of engineering time, shorten schedules, reduce errors, and simplify start-up. (More about specific useability and documentation considerations in Chapter 7.)

Improvements in communications

The major improvement in communications is the trend away from analog signals, which not only tend to drift, but which also require dedicated pneumatic tubing or a twisted pair of wires per signal. New systems today use digital signals that can be batch-transmitted, or "multiplexed." Using a multiplexed digital communications bus allows significant reduction in wiring, plus faster system response. Fiber-optic technology is now replacing wire cabling in many applications because it provides unparalleled speed and noise immunity, plus excellent message security and long signal ranges.

Fiber-optic transmission rates can exceed hard-wired copper cable transmission rates by a factor or 10 or more. In addition, fiber-optics pose no electrical hazard, and the signal cannot be disturbed by induced voltages from adjacent electrical activity (motors surging at start-up, etc.).

"Smart" transmitters also make possible improvements in communications: Instead of transmitting data for analysis at some later time in some distant location, they can carry out the first analysis on the spot. If a sensed value is within normal limits, it may not be transmitted at all.

This kind of efficiency—exception reporting—is very important in speeding data analysis and alarm processing within the individual process control plant loop. It means that critical data is handled immediately without waiting for a perhaps prohibitively long time to elapse before action is taken. While the exception-reporting feature guarantees faster response, it also guarantees that no signal will go untransmitted for longer than a set maximum scan period. Detailed tests and analyses show that exception-based systems are more responsive than simple scanned sensors and have a bandwidth more than 20 times greater.

New software techniques like advanced data compression are also speeding plant-loop communications and expanding the opportunities for even more comprehensive systems.

Levels of automation: a conceptual approach

Now it's time to actually come to grips with automation systems themselves. We'll lay down a conceptual framework around which we can interweave all the applications, hardware, and software that go into making automation work.

The three lower levels of the column are all real-time systems geared to the events transpiring every second. The top two levels are today frequently not real-time systems, but analytical and planning tools used periodically to achieve profitable operation of the business. All of the bottom four levels of automation work together as part of the strategic process management of the

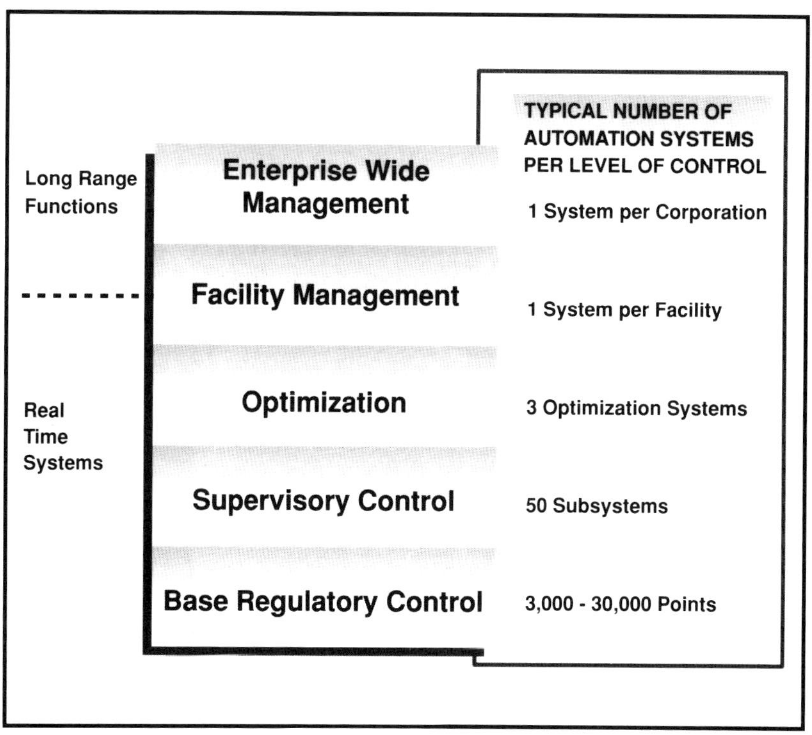

Figure 4-6. Levels of automation within a single platform

platform. We'll review briefly what happens at each level here (subsequent chapters will focus in more detail), starting from simple base-regulatory and supervisory functions to the statistical process control and artificial intelligence applications that play such an important role in optimization and facility management.

Level 1: Base-regulatory level

The bottom of the column is the Base-regulatory level. Automation functions at this level deal with the sensors and actuators for a single piece of equipment or for a small cluster of devices. Key functions include:

- *Sensing of variables.* Four of the most common variables at the process level are temperature, pressure, flow, and level.
- *Initial conversion of data to information.* This includes scaling, linearization, filtering, conversion to engineering units, and perhaps limit checks. The devices that may be involved include older pneumatic and electronic analog devices (such as flow computers), single-loop controllers, and the "smart" or "wise" transmitters that are gaining increasing acceptance in the market.
- *Communication* to the next level of the system. This may take place via pneumatic tubes, twisted-pair copper cable, fiber-optic cable, or via direct microwave, radio, or satellite transmission.

Our example in Figure 4-6 illustrates a platform with 3,000 points at the base-regulatory level, measuring temperature, pressure, flow, and level, and controlling valves and switches.

Level 2: Supervisory level

At the second level from the bottom, we have computing and data management functions going on in some format or another. They may be in an older central process-control computer, in a Personal Computer running SCADA (Supervisory Control and Data Acquisition) software, or, in more modern systems, integrated with microprocessor applications in the racks of a distributed control system (DCS). Some typical functions that take place at the supervisory level include:

- *Automated startup and shutdown* of equipment—pumps, compressors, backup generators, etc.
- *Logical operation of a process* or processes. The system analyzes inputs and activates outputs to keep processes running as specified. No operator intervention is needed.
- *Operator interface.* The operator interface allows humans to communicate to the system. The mechanism may be a 4 by 6 inch pushbutton panel next to the process, a hand-held plug-in controller, one or more CRT screens with keyboards or touch-screens, or, in older designs, custom-built panelboards full of lights and buttons. The supervisory system must also have some mechanism for loading the process-control logic.

  In the case of a central computer or PC-based system, programming may be done on the system itself.

  Rack-mounted distributed systems generally include an advanced integrated engineering workstation, used to create new control configurations and down-load them to the control system itself. Engineering workstations can also be used to diagnose problems and insure the "health" of both the process and the system.
- *Alarming systems.* For simple problems, notifying an operator of an alarm condition is sufficient. But as platform processes and supervisory systems become more complex, potential problems may become more severe. In many cases, it's helpful to incorporate an "expert" alarming capability that prioritizes alarms so the operator can deal with the most significant alarms first. Many such systems even provide guidance as to what to do next.
- *Equipment diagnostics.* This is where some of the new technologies that we talked about in the previous section come into play. The supervisory system manages the diagnostic information about process equipment (for example,

vibration monitoring might indicate that the bearings on a mud pump need replacing), and it also manages diagnostic information about itself, the control system. It should be able to diagnose, report, and recommend solutions to both its own problems and those of the process.

- *Fault tolerance.* Typically, a supervisory system should not only be able to diagnose itself, it should be invulnerable to the vast majority of failures and able to switch to a backup to keep a things running. Ideally, the system will isolate faults so exactly it will tell the operator which module to change out. Advanced systems today carry that level of sophisticated "self-repair" right down to the "smart" or "wise" transmitter in the plant.
- *Automatic remote tuning.* Advanced supervisory systems can tune themselves (resetting span and zero on "smart" transmitters and resetting conversion factors) in response to production levels, weather conditions, or other operational parameters.
- *Data and event logging.* The supervisory system should keep track of all alarm conditions, as well as pertinent operator actions. In addition, it may also maintain a database for applications like predictive maintenance. In order for the next level of the pyramid to work, the supervisory system must be able to maintain data used by optimization systems.

The example in Figure 4-6 shows the same platform that had 3,000 control points having 50 supervisory systems. These systems might include:

- *Process systems:* Wellhead, gas lift, separation, oil treating, water treating, compressor control, reservoir management, etc.
- *Safety and Emergency Shut-Down (ESD) systems:* fire control, gas detection, emergency shut-down, structural analysis, ballasting, etc.
- *Support systems:* Power generation, HVAC, fresh water treating, waste-water treating, etc.

Level 3: Optimization

While supervisory systems can be made to operate in rack-mounted microprocessing modules that don't require major operator intervention, optimization systems generally require full-blown computing capabilities and more operator involvement.

In the past, even relatively simple applications (by today's standards) required expensive high-performance mini-computers. Today, the same applications are being absorbed by digital distributed systems, as microprocessors approach (or exceed) the capabilities of the most powerful supercomputers of the early 1980s.

In fact, even supercomputers today are being constructed of micro-processor arrays linked by super-speed communications. Such systems are remarkably similar in structure to current advanced distributed control systems.

The roll of the operator in an optimized system is to serve as a fairly high-level process manager, using to the fullest extent possible the operating "leverage" provided by the system.

Optimization covers the integration of related supervisory systems, and it requires communication between systems. Typical examples of optimization are:

- *Upstream/downstream integration* of the process stream. While each process on a production platform has its own parameters and its own supervisory control system, most processes are related to each other in some way or another. The separation process depends on the results in the test separator, the water-treating system depends on separation, and so on. Optimization systems allow multiple process systems to work together most efficiently.
- *Coordination of related applications.* For example, a fire detection system might be tied in with the HVAC system, the platform information database, and the platform's public address system. A 500° grease fire in the kitchen could trigger the appropriate extinguishant system, notify personnel over

Figure 4-7. A very large complex control system complete with embedded artificial intelligence and large wall monitor displays

the PA (using computerized voice synthesis), and alter air conditioning to cut off air to the fire while feeding fresh air to crew quarters. A 2000° fire in the separation unit could trigger platform evacuation. The decision to evacuate a platform can involve so many factors that it may be that only an expert system can arrive at an accurate answer fast enough to do any good.

- *Imbedded expert systems.* By including an on-line knowledge database and the appropriate expert systems in real-time mode, the optimization system can make complex decisions in seconds that humans might take hours to reach. The fire and safety example above is a good example of an expert system at work. We'll have much more to say about expert systems in the next chapter.

- *Sharing of facilities by different systems.* For example, most platforms have two water-treating systems, one for process water and one for the human support system. Clearly, the requirements of each system are quite different, although they may be able to share certain facilities.

- *On-line statistical process control.* SPC is a proven technique for controlling processes so they that operate consistently within specifications. In the past, SPC has been primarily an off-line evaluation tool, but today's systems allow SPC to be used as an on-line tool interfacing directly with the supervisory control system to keep processes working optimally. We'll discuss SPC in depth in the next chapter.

The column in Figure 4-6 shows three optimization systems. They might be:

- All process systems;
- Safety and ESD (emergency shutdown) systems; and
- Support systems.

Level 4: Facility management

Below this level, real-time systems monitor what is happening at any given moment. A facility management system looks to the future and incorporates non-real-time functions along with knowledge of external factors, helping owners and operators to implement management strategy and define their operating tactics. Facility management can cover all aspects of an offshore platform or group of related platforms.

Facility management applications relate directly to on-line supervisory control functions, or they may be separate decision-making systems. Examples of facility management functions are:

- *Maintenance scheduling.* We have already mentioned vibration monitoring, and that is a good example of direct input to facility management. As a supervisory system, vibration monitoring can automatically shut down a high-speed gas turbine whose bearings show signs of wear to prevent the kind of explosion that takes planes out of the sky. By the same token, the vibration monitoring system can keep a rickety mud pump working at its usual 25 rpm until routine maintenance can fix it. By logging incipient maintenance problems to a database, the vibration monitoring system provides the information to predict "end-of-life" and prioritize maintenance, an important function of off-line facility management.

- *People and materials scheduling.* If you know you are scheduled for a major shutdown and rework of a well, you may have 20 contract people on the platform in two weeks for 10 days, which can amount to 600 meals.

- *Off-line statistical process control.* Evaluation of data collected by the on-line system can lead to redesign and other improvements in process systems.

- *Off-line expert systems.* Using Pareto Analysis and other techniques, evaluation of faults and decision-making patterns can help focus attention on the 20% of problems that lead to 80% of the expense and lost time.

Figure 4-8. A typical large drilling and production platform

Level 5: Enterprise-wide management information systems

Enterprise management relates most heavily to the financial side of the business. For example, fluctuation in the price of crude affects platform operation: If the price drops it may be most economic to shut the platform down, get people off, and save as much as you can. If it appears that the fluctuation is temporary, it may be better to curtail production and go to a skeleton crew temporarily.

From a liability perspective, insurance rates are often based on the number of people on a platform. Using input from platform facility management systems, the corporate risk manager might be able to prove that although platform is insured for 100 people, for 75% of the time only 40 people are present. This could reduce liability costs by a proportionate amount.

While these applications are undeniably important, they fall outside the purview of this book.

The role of mainframes, PCs, and distributed control systems

As we've talked about levels of automation, we've worked with a conceptual framework that is separate from the equipment where the actual applications reside. The important point to be made by examining all levels of automation possibilities is that these are genuine possibilities. The equipment we choose for the implementation of automation systems should be based on an architecture that will support the future that lies ahead of us. Obviously, no one starts out with the whole thing, any more than Henry Ford's first car was a Lincoln Continental. But today we have a good idea of future directions, and we should be prepared to move that way.

So we will examine now three primary trends in automation systems with an eye toward which will best help us capitalize on the opportunities the future has to offer.

Mainframe computers

Mainframe process control systems have been around since the early days of computing systems. The IBM 1800 system in the early 1970s was designed specifically for process control, while later IBM general purpose computers (the 370 line in the mid to late 1970s) were used in IBM's Advanced Control System (ACS), initially marketed to reineries and chemical plants.

Digital Equipment's early efforts, the PDP-8 and PDP-11 series, were among the first steps taken by a major computer vendor to offer a platform for direct digital process-control (DDC) systems. These systems performed well in on-shore situations, but their size, complexity, and requirement for special environments and programming made them considerably less successful offshore. Today, Digital's micro-VAX line of computers indicates a strong commitment to the process industry; as computer systems become more rugged, smaller, and ever more powerful, they should find their way offshore for higher-level facility management and even enterprise-wide management functions.

While mainframes appear rarely if ever in offshore operations, they do offer several apparent advantages:

- They can **communicate** with older PLCs, supervisory systems, and distributed systems, although this may require special programming and communication hardware;
- They have the number-crunching and database **computer power** it takes for matrix calculations, linear programming, and complex statistical and expert systems; and
- They may have a **ready-made interface** to existing high-level management information systems (MIS).

But mainframes also have several critical disadvantages that have limited their application to process control, especially in the offshore environment:

- Originally, mainframe computers could only do one thing at a time, operating as **serial processors** to handle each process variable or operator message as it came in. This posed a problem, because as operations get more complex, the number of variables that must be scanned increases, and the computer may be unable to respond fast enough.

  Since the early 1970s many systems have attempted to circumvent this problem through multitasking operating systems. While this type of operating system can handle more than one job at a time, it doesn't make the basic architecture any faster. Newer approaches include the use of multiple processing units within a mainframe computer, but this tends to be both complex and expensive. It is simpler and cheaper to gain true multiprocessing capability by using a distributed system.

- A mainframe is reliable on its own terms, but process designers must remember that **"all the eggs are in one basket."** Many industries that led the way in process control applications started out with mainframe or large minicomputers and now use distributed systems specifically to solve reliability problems. Control applications in refining and paper-making are two examples.

- **Communications hardware can get out of hand.** As you can imagine, the idea of every process variable having a twisted pair to the back of the mainframe quickly becomes ridiculous. Data-multiplexing "slaves" can be used; this solves the physical wiring problem. Such slaves are powerful enough to assume some of the processing duties, so the mainframe computer metamorphoses into the hub of a (not-too-attractive) distributed system.

- Mainframes **cannot handle remote process input/output (I/O)** very well, because they require data concentrators and communication processors to accumulate, format and present the data in predigested form. These preprocessors are microprocessors or even full-blown computers in their own right. And we're back to a not-too-terrific distributed system.

- Mainframes require a high level of **abstruse computer expertise.** That means programmers and systems people. Most mainframe process control programming involves code written in a standard computer language like Fortran or "C." The complex logic and mathematical algorithms of these languages may require alterations every time some component of the process is changed. Depending on the modularity of the code, this can be range from an annoying problem to a nightmare.

  In addition to computer expertise, mainframe computers often require special physical facilities, an air-conditioned environment, and a staff of maintenance personnel—all of which offset the space, weight, and staffing advantages that were one reason to automate in the first place.

The result of these factors is that mainframes and mini-computers (which have taken over many of the roles of mainframes) are often simply not a practical, reliable, or cost-effective solution to our automation problems. This becomes increasingly true when we take into account the growing power and flexibility of distributed systems.

Personal computers

Personal computers have invaded the world, and the offshore industry is no exception. While mainframes have taken a top-down approach to the automation conundrum, PCs have attacked it from the bottom up. While mainframes like to see things globally, PCs like to look at individual applications.

Increasingly, PC-based SCADA (Supervisory Control And Data Acquisition) systems are being used to control both single and multiple processes. DOS-based systems are generally limited to one task (one SCADA program) and one operator console at a time, but newer UNIX-based systems offer multitasking, multiple operator consoles, and a greatly expanded number of applications. As personal computers continue to become more powerful, their use in supporting process control systems will doubtless continue to expand. PC computers have many advantages:

- *Low cost.* For well-defined individual applications that do not require on-line SPC or expert systems, the PC may be the ideal solution. Software is available today for a wide range of applications and I/O devices.
- *On-site availability.* Because the PC handles a relatively small number of functions, it can be located close to the process. That reduces wiring costs and makes the unit accessible for whatever monitoring and operator intervention may be required.
- *Communications with other PCs, mainframes, and minis.* Networking systems allow PCs to share data with other PCs on a local area network (LAN) like Ethernet. PCs can also upload and download from mainframes and minis.

Unfortunately, PCs have some disadvantages, too:

- *Low-level processing.* PCs perform best at the base-regulatory and supervisory level. You may have trouble integrating applications due to the processing limitations of any single PC.
- *Size.* We have just said that PCs may not be big enough, but in some instances, they may be too big. Dedicating a PC to a simple isolated process where a single-loop controller could do the job is probably uneconomical.
- *No on-line SPC or expert systems.* PC-based SPC and expert systems are not well developed as we enter the 1990's and are, to our knowledge, still off-line systems. We know of no currently available systems for on-line SPC and embedded on-line expert systems.
- *Software limitations.* Software for different PC-SCADA systems handling different processes may be incompatible with each other, making it incumbent upon the user to maintain system standards and overcome programming incompatibilities.

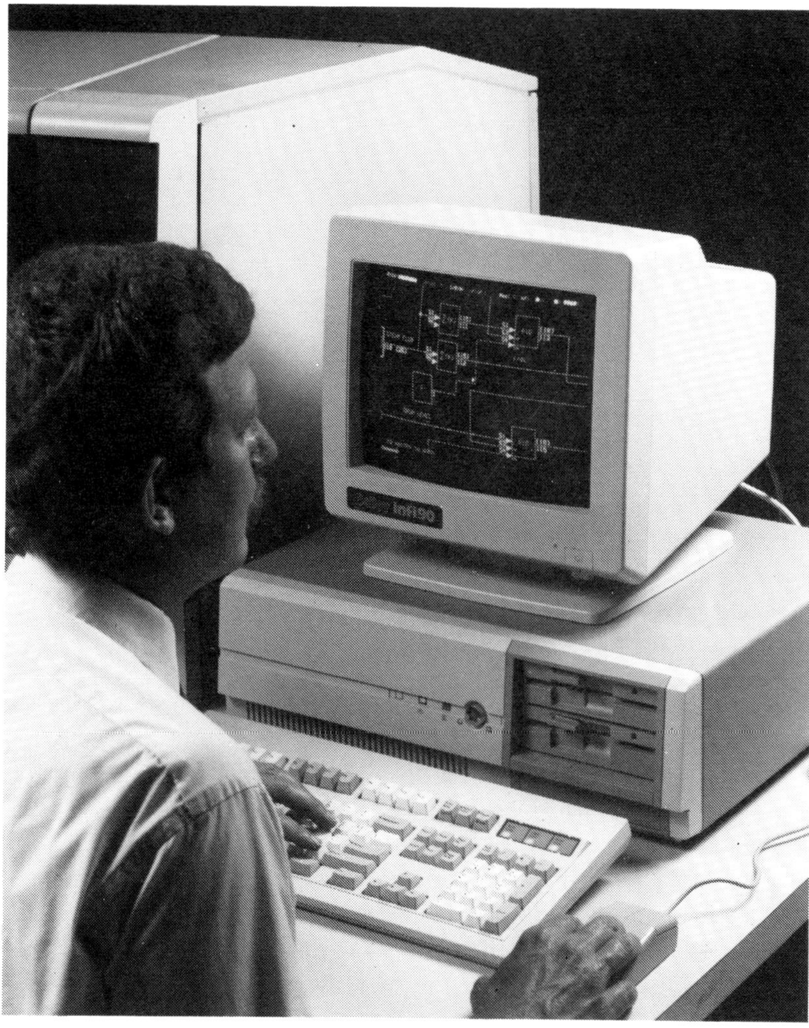

Figure 4-9. A small personal computer based control system

Although multiple PC-based systems can be structured together to provide a crude form of DCS, the lack of integration in the current state of the art and the limited flexibility in the size of the units may prevent users from taking advantage of the full range of possibilities that automation provides.

An even more critical consideration is the lack of environmental hardiness designed into the hardware, software, and communications capabilities of these systems. Their development in a benign scientific and commercial world may make them ultimately unfit even for the rigorous demands of critical onshore industrial applications, much less offshore systems.

Distributed control systems

Distributed control systems (DCS) seem to offer the best combination of the advantages of mainframe and PC-based automation. They are composed of multiple rack-mounted modules, operator interface stations, and (optional) processing and networking computers. Modules may have whatever size and function is appropriate to the job at hand.

Distributed control systems offer many advantages:

- *Flexibility*. Because individual functions are isolated on separate modules that can be combined in dozens or perhaps hundreds of different ways, a DCS offers considerably more flexibility than either a mainframe or a host of PCs. A DCS can cost-effectively implement base-regulatory and supervisory functions that would leave a computer vastly under-utilized, and it can combine enough of these capabilities to equal or exceed the processing power of a mainframe or even a supercomputer.

- *Parallel operation*. Unlike a mainframe, which typically operates in a serial/sequential manner (although as we noted previously, recent mainframes also use 2-, 4-, 6-, or even 8-way multiprocessing), a DCS operates in parallel: Each element of the system does its own job and communicates to other elements when necessary. A well-designed DCS won't succumb to "token choke," because the system's architechture works to allow every component on the data highway to send out a message simultaneously, without failures or lost messages. Some systems can simultaneously carry up to 250 or more messages on one communications loop. In addition, each loop can have subloops—in some systems, as many as 250 or more. Such a system can handle up to 250 x 250 messages at the same time.

If a plant upset or accident occurs, the DCS system architecture easily supports all the alarms that come in. This architecture, combined with high-speed, secure communications and lightning-fast processors for alarm processing, can handle even the most time- and mission-critical applications with performance that exceeds any other type of system. Theoretically, the size of today's most modern systems is almost unlimited.

- *Works well with "smart" and "wise" transmitters.* These new I/O device interfaces are specifically designed to convert signals to a digital format that can be multiplexed on a single twisted pair. This has real advantages: It offloads scaling, linearization, filtering, and other functions from the processing and logic modules, and it can drastically reduce the amount of wiring needed.
- *Easy programming.* DCS systems typically come with a "built in" control and engineering language, rather than requiring programming as mainframe computers do. This means that control and process engineers without programming background can quickly learn to configure and reconfigure the process control system. Because these "firmware" systems are interpretive, a change to one part of a control strategy does not require reprogramming. A DCS, as a result, is much easier to install and start up, and much easier to change and maintain than either conventional analog and sequential logic systems or computer-based control systems.

Of course, a DCS can have some disadvantages, too:

- *Potential software incompatability.* Because these are special-purpose applications, they rely on vendor-supplied software. Potential users should check out such factors as ease of configuration and configuration changes, documentation, availability of SPC and expert systems, and so on.

- *Quixotic vendor support.* While these systems offer the greatest potential expandability, the future with any one of them is only as solid as the vendor and the vendor's vision and commitment to the future.

Who does what

As we might expect, mainframes and minicomputers are most effective at the top of the column, although some vendors are trying (with varying degrees of success) to spread the effective range of minicomputers downward into the supervisory level.

PCs start from the bottom, although as we have mentioned, they may be an inappropriate overkill at the lowest base-

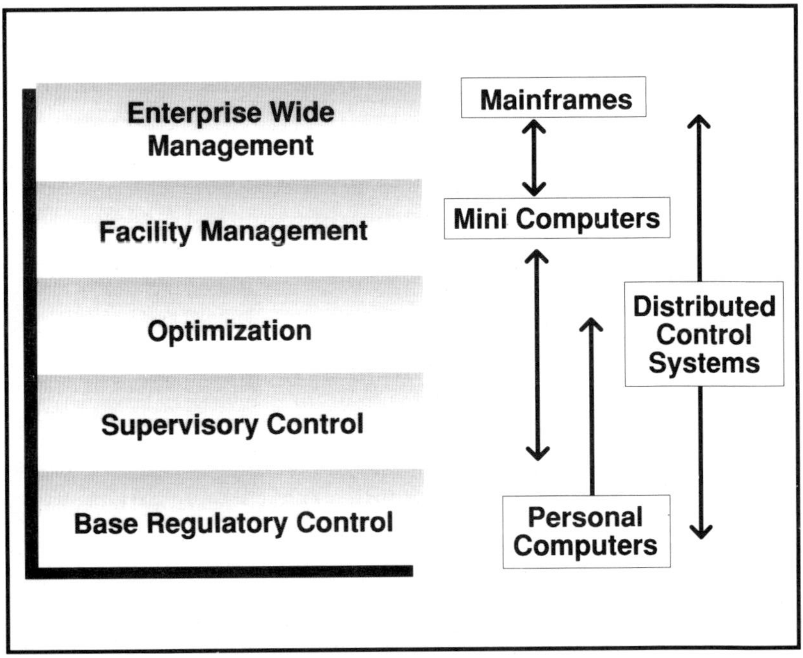

Figure 4-10. Domains of control

Figure 4-11. A large control room

regulatory levels. As PC evolution continues, they push higher up the pyramid, displacing minicomputers and even mainframe functionality.

Distributed control systems seem to offer the most comprehensive approach because of their modular nature; however, the more sophisticated DCS optimization and facility management applications are only now being exploited.

Due to rapid advancements in technology, the boundaries of the middle levels of control are rather ill-defined and in continuing flux. The future may clarify the situation—or it may bring us yet unthought-of solutions.

Summary

In this chapter, we've defined automation as a means of executing management strategy and tactics, an approach to management that can reach into nearly every area of offshore platform operations and management. We've described process control applications as a natural extension of the kind of things humans do all the time—perceive a situation, analyze it, and take action. And we've briefly discussed some of the technologies that are driving automation at a faster pace today than ever before.

Of course, the main emphasis of this chapter has been to structure and define the environment of platform automation and the types of systems available to work within that environment. The industry is changing so rapidly, that often events and advancements seem to have no clear relationship to one another.

Chapter 5

COMPONENTS OF STRATEGIC PROCESS AUTOMATION

What is strategic process automation?

The idea behind strategic process automation is the integration of business decision-making with actual process control, so that information and informed action flow seamlessly within the organization.

To make strategic process automation possible, we must take an integrated approach to all four of the lower levels of the columns in Chapter 4—capturing data and information from the base-regulatory level, folding that into supervisory control systems, optimizing those supervisory systems to work together, and using all the knowledge we gain from optimization to produce the best economics for the platform. In a strategic process automation approach, scheduling, maintenance planning, forecasting, and many other management issues become a natural extension of the process control system.

The importance of integration

Strategic process automation implies a high degree of integration, and, indeed, the importance of integration cannot be overstressed. An integrated approach brings with it:

- *The free transfer of data, information, knowledge, and wisdom.* This is the key benefit and a primary ingredient for more profitable operation.

- *Commonality of equipment.* With one type of equipment used throughout, you gain expertise quickly and spend your time implementing applications instead of learning, interconnecting and maintaining new systems.
- *Simplified support.* You learn one set of jargon, one set of engineering and design tools, and one set of operational standards.
- *Elimination of operator confusion and errors.* Operators become familiar with how the one system works, and don't have to learn variations in the ways that differing approaches handle similar problems. As the Three-Mile Island nuclear facility tragedy in the U.S. demonstrated on-shore, this can become critical.
- *Elimination of operator confusion and errors.* Operators become familiar with how the one system works, and don't have to learn variations in the ways that differing approaches handle similar problems. As the Three-Mile Island nuclear facility tragedy in the U.S. demonstrated on-shore, this can become critical.

In one offshore example, a platform had a fairly integrated DCS and a separate Fire, Gas and Safety (FGS) system. Although the fire and gas information was fed into the DCS and available on the DCS CRTs, fire fighters were not familiar with the DCS. When a fire occured, it appeared as a light in a general area on the FGS system panelboard, and the fire brigade ended up at the wrong location. While they hunted for the fire, considerable damage occured. Had the fire personnel known how to use the DCS, or had basic ergonomic principles been used to interface the FGS functions with the DCS, the fire fighters would have found the *exact* location of the fire and put it out. The best solution would have been implementation of FGS functions in DCS hardware with appropriate partitioning and secure communications, which could have prevented the problem altogether.

- *Simplified communications.* It is easy to communicate between and within all levels of an integrated system. Bear in mind, however, that although integrated systems are designed to communicate among all their own parts, they may not talk to other systems. For example, Company A's control systems may talk to their own "smart" and "wise" transmitters, but not to Brand B or C, and the reverse may also hold true. This becomes an important consideration once you fully understand all the advantages "smart" and "wise" transmitters have to offer. While we can bemoan the current lack of standardization, it still behooves us to be aware of constraints like this early in the design phase. When you choose a system, consider not only the current status but also the vendor's commitment to support ongoing technical enhancements of their equipment, including retrofitting to standards as they emerge. Such an approach is called "technological transparency."
- *Elimination or minimization of interfaces.* Chances are, every time you connect Brand X to Brand Y, you or the vendor(s) will end up designing and implementing some interface. Every nonstandard interface is a problem waiting to surface. Even if it works flawlessly today, it may not function with the X and Y technology of tomorrow. Again, look for technological transparency: ease of interface and adherance to interface standards. Whenever possible, integrate functions directly into the DCS.
- *Increased reliability.* An integrated system means that you can pinpoint responsibility to one vendor, which by itself can make a drastic improvement in reliability. And again, you don't have the inherent problems of do-it-yourself interfaces to deal with.
- *A single spare parts pool.* Obvious, but why is it so often overlooked?

One reason fully integrated systems are so scarce today is that many facilities are designed by architectural and engineering firms whose primary incentive may be lowest aggregate initial purchase price. The real benefits of integrated automation don't show up until installation and commissioning, and they really don't blossom until operations are under way. During the purchasing phase, integrated, strategic, process-control automation may appear to be needless expense.

Regrettably, architect and engineering firms with less experience in process control and plant automation and optimization may use design engineers who have only a limited familiarity with processes, much less process-control systems. They plan for rudimentary controls at the base-regulatory level, but have no real understanding that it is through the overall optimization of the processes and base-regulatory functions that you can achieve good plant operability and maximize process economics.

If you're an owner, get involved. Bring process engineers into the design phase along with those hardware design engineers. It is much easier to do it right the first time than it is to retrofit. Paradoxically, and unfortunately, it is often easier to justify a retrofit when personnel get tired of living with operating nightmares.

In line with the importance of integration are the benefits to be gained by using true distributed control systems. A DCS is not put together in a hierarchical, pyramid-like manner: Yet it has the greatest capability to realize the opportunities at all levels of our automation pyramid, because it can do so many things at once in a consistent, coordinated, timely manner.

Elements of control within a distributed control system

- *Redundant communications* are an essential part of any mission-critical system. They allow various parts of the system—process control subsystems, operator interface stations, and computers—to communicate with each other in a flexible way that is easy to upgrade when the need arises. Additional process control subsystems, operator interfaces, and computers can be added to each communications loop as needed. In addition, the loop can communicate to other loops and systems through local and remote communication interfaces.

- The process control subsystem consists of the modules shown within the dotted lines on Figure 5-1: the processing logic, device interfaces (slaves), and simple operator interfaces. Starting from the bottom up:
  - The **termination units (TUs)** accept direct connections from analog and digital input/output (I/O) devices—the

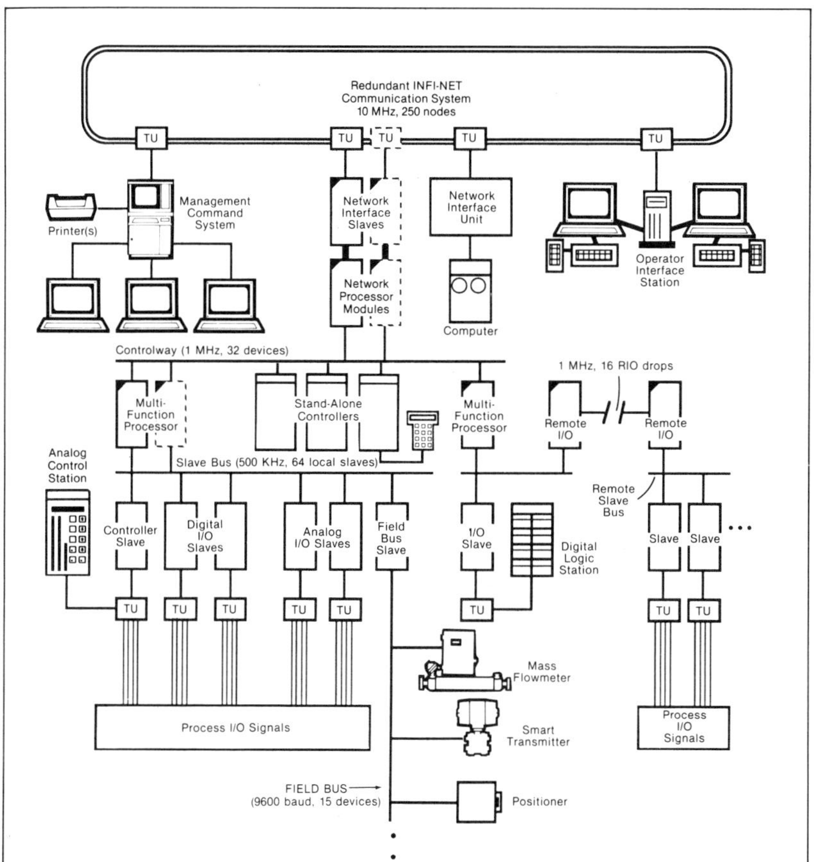

Figure 5-1. Digital Control System schematic

devices that measure temperature, pressure, etc., and control valves and switches—and pass these signals on to analog or digital cards.

- The **analog and digital cards** convert data to digital values and perform basic functions like signal quality checking, linearization, scaling, conversion, and filtering.
- The **operator interface card** accepts operator-entered data from mini-panelboards or hand-held plug-in controllers. This is a simple form of interface that allows operators to recalibrate instruments and change set-points.
- The **field bus slave** handles I/O on a multiplexed single twisted pair that serves multiple "smart" and "wise" transmitters and receivers (both sensors and actuators).
- The **multifunction processor modules** perform further input signal processing, execute basic and advanced control logic, and send output signals back to output cards. Along with optional additional logic modules, this module has most of the logic capacity of the process control subsystem.

As process control becomes more complex, the process-control subsystem can use the communications loop to integrate several subsystems and to provide more and more sophisticated operator interfaces and computers, as needed.

- **The operator interface station** provides high-level operator interface units usually featuring one or more CRT screens with keyboards designed specifically by human-factors specialists for operator use. Today the trend is definitely away from panelboards and toward greater use of the CRT-based console because of the greater flexibility that CRTs offer. In some situations, transition designs are present, and panelboards and CRTs complement each other: One safety system uses a map panelboard to indicate the presence of gas in various parts of the plant. A quick look at the lights on this board indicates where a cloud of gas is and how fast it is moving. CRT screens then show the exact location of leaks and give instructions for damage control. An alternative to this approach is to use high-resolution projection CRTs to integrate both panelboard and CRT functions.

- **The computer interface unit** provides on-line and off-line computing power to the system. This unit makes it possible to tie into the processing power that may be needed for reservoir calculations, maintenance scheduling, and other high-level functions. It can be a PC, minicomputer, or mainframe, and it can have its own connection to a Local Area Network (LAN). It can serve as a sort of "Management Command System" for the DCS.

- **The engineering workstation** is the tool that makes it easy to get a DCS on the air. It is used for:

  - The design of process control strategies using CAD, engineering ladder logic, "C" programming, etc.;
  - Maintenance of the database(s);
  - Reporting for scheduling and maintenance functions;
  - Creation of expert systems and SPC analysis tools; and
  - System utilities (backup, etc.).

- **The local network interface** provides access to other communication loops on the same platform. This is a cabled interface that supports the exchange of real-time data and makes the distributed control system capable of being expanded to almost any size.

- **The remote network interface** provides access to other communications loops that are located too far away to run cable. This could be another platform or a shore location. Access here is via radio-frequency, microwave or relayed-satellite communications. Microwave is acceptable for short distances with a clear line of sight where good weather prevails. Satellite communication is becoming the method of choice, because it doesn't have the interference problems that microwave does.

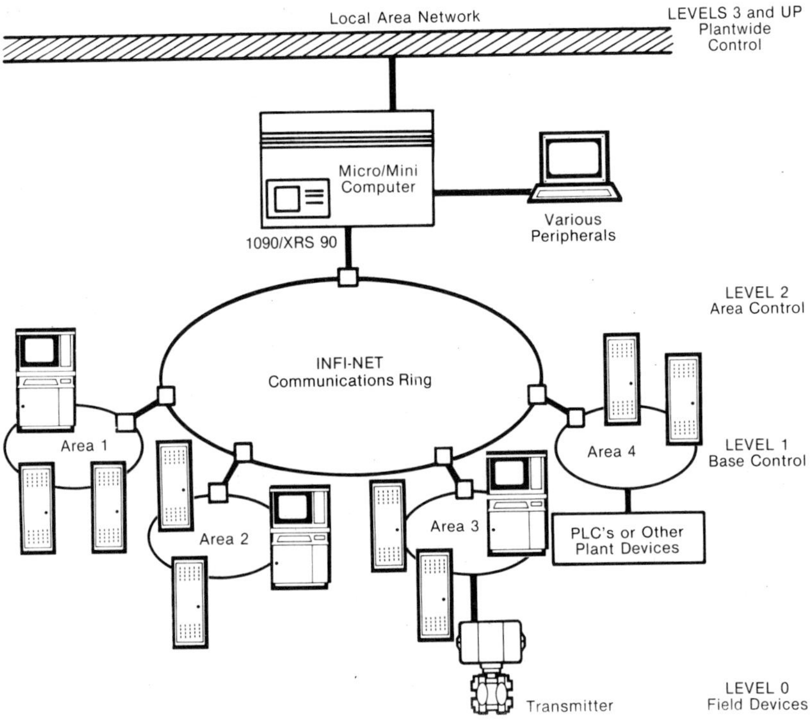

Figure 5-2. Multiple loop configuration example

As you can see, the power of distributed architectures lies in their ability to accomodate and integrate a mind-numbing variety of previously separate and often inconsistent functions. Small systems might have only a few components of a process control subsystem, while larger systems might have several process control subsystems and more complex operator interfaces with CRT screens and printers. Still more complex systems might also include extensive imbedded computing facilities for such applications as on-line SPC and expert systems. The most highly developed systems might involve several communications loops, each with its own processing units, computers, operator interface stations, and management command system. This type of configuration (shown in Figure 5-2) offers the possibility of unified control of the entire facility with all its applications. In subsequent chapters, we will go into detail about the functions that take place within distributed control systems. Chapter 6 will cover data acquisition and control considerations for devices, pneumatics, programmable logic controllers, "smart" and "wise" transmitters, and retrofit applications. Chapter 7 will look at process control considerations, the factors relating to the supervisory control level, particularly fault tolerance (dual and triple redundancy), diagnostics, alarm prioritization, maintenance, useability, and documentation. We will also discuss safety systems. Chapter 8 will cover facility management considerations, focusing on SPC and expert systems.

Chapter 6

DATA ACQUISITION AND CONTROL CONSIDERATIONS

This chapter starts at the lowest level of automation: the sensors and measurement transmitters for temperature, pressure, level and flow. It then moves up a level to the various techniques for using the measured data. We'll discuss pneumatic systems, programmable logic controllers, conversion considerations for migrating to a Distributed Control System, communication technologies, "smart" transmitters, and the operator interface. Our intention is to provide a good overview of the important components of automated systems.

A quick look at accuracy, precision, repeatability, hysteresis, and random errors

Control can only be as good as the quality of measurement. If the techniques used to transport measured data to the logic areas of the system degrade the original measurement, it limits the ability to accurately control. Thus any discussion of measurement techniques should be based on an understanding of the concepts and terminology of accuracy, precision, and repeatability.

Accuracy

Accuracy is a term that presumes knowledge of the "true" value of a variable against which the measured value can be compared. Accuracy is similar to measuring how close you are to hitting the exact center of a bulls-eye whose position is clearly known.

Accuracy can be stated in terms of deviation from a defined point, like the freezing point of water at a standard pressure; or from a mathematically described curve connecting two such defined points; or from a table of empirically derived points on a curve.

True accuracy is rarely the objective in industrial control measurements, and the term is often misused.

Precision

Precision is a term that relates to the resolution of the scale used to make a measurement. In other words, accuracy can be measured with a precision stated per inch or per thousandth of an inch. Inaccurate measurements, perhaps displaced from the true bulls-eye by several inches in any direction, can still be measured with a precision of thousandths of an inch.

Repeatability

Repeatability is a special characteristic of measurements that depends on some inherent fixed offset from bulls-eye accuracy. If a measurement has good repeatability, you achieve control by adding an offsetting constant value to the measurement to bring it back to an accurate representation of the value measured.

A measurement and control system that provides repeatability with sufficient precision—rather than true accuracy—is generally the practical ideal in industrial automation systems. Unfortunately, this is usually an oversimplification of reality.

Nonlinearities

Real-world measurement techniques usually suffer problems due to nonlinearities, either in the measuring scale or in subsequent mechanical or electronic devices in the data acquisition system.

Some nonlinearities are inherent in the measurement principle itself—electric power measurements and liquid flow rate measurements, for example, are related to the square of the electric current and the square root of the pressure drop, respectively.

Other common nonlinearities are due to saturation or integration effects and limiting effects. A capacitor is a good example of an integrating or saturating effect—as it charges, its back-voltage increases to effectively reduce the charging voltage. A spring is a good example of a limiting effect—it can only be compressed so far before "bottoming out".

Secondary but important nonlinearities also result from environmental sensitivities of the measuring device. As temperature increases, for example, a metal ruler grows in length; at a given forward voltage, the current through a diode increases.

Nonlinearities require compensation of the output produced by the measuring device. Straight-line functions are easiest to compensate accurately. Curvilinear functions such as square and square root are much more difficult to compensate.

Hysteresis, deadbands, and backlash

Hysteresis is the name for a complex nonlinearity function that occurs in magnetic materials, dielectric materials, and mechanical springs. Generally, hysteresis deforms the straight or curved line input/output relationship of the measuring device into a loop, so that the curve for an decreasing input is displaced from that for an increasing input.

Hysteresis takes place when there is an energy storage mechanism that requires the input to be reversed through some minimum amount before the output reverses. The "lag" between input and output values due to hysteresis is generally not a time-sensitive function, at least not at industrial control-system bandwidths.

Other nonlinearities can occur in measurement devices that cause a loop-like curve that resembles hysteresis, such as deadband and backlash. These can vary with time within control system bandwidths, which compounds the difficulty of precise compensation. Hysteresis, deadbands, and backlash should be avoided in measurement devices and data acquisition systems.

Random errors

Finally, there are random errors, such as those due to noise on an analog signal. Quantization errors due to analog-to-digital conversions, also have a pseudo-random appearance.

All measurement and data acquisition system nonlinearities must either be avoided or accurately compensated for to achieve repeatable, precise control.

Basic devices used to acquire physical data

Four process variables account for most of the measurements made in process control: temperature, pressure, level, and flow. These four measurements, in combination, give the control and automation system a substantial portion of the information it uses to make decisions about control actions.

Other measurements, while less frequently used, also have substantial impact on control systems. Among the devices used to make these more exotic measurements are:

- Analysis instruments that determine the composition of products;
- Vibration and displacement sensors used in machine condition monitoring;
- Wind, wave, and other environmental monitors;
- Fire and gas detection monitors;
- Strain gauges, acoustic emission systems, fault detection monitors and other devices used to determine structural integrity;
- Density, viscosity and other flowing stream monitors.

This latter group of measurements directly address the issues that are central to advanced automation and total facility management systems: product and intermediate stream composition and quality, operability, safety, reliability, and profitability.

Before we consider the more exotic devices, a basic review of the "frequent four" measurements used in all process control systems is in order.

Temperature sensors

Temperature measurement has taken many forms over the years. Simple bulb thermometers are still used, but in general they are neither rugged enough nor accurate enough for process control service. Today, the most common industrial temperature sensors are thermocouples and resistance temperature devices. Two other often-seen temperature sensing devices are thermistors and dual-frequency optical pyrometers.

Thermocouples

A thermocouple is a device formed by a junction or coupling of two dissimilar metals. They are inherently small and rugged and have reasonably good long-term stability, making them well-suited for use in industrial environments.

At low temperatures where the atmosphere is clean, bare junctions can be used. Most industrial thermocouples, however, are protected by sheaths of materials resistant to atmospheric corrosives as well as high temperatures. Ceramic sheaths inside metal sheaths are used at very high temperatures. Any sheath introduces time lags in the temperature measurement.

Thermocouples depend primarily on two physical principles for their operation: the Volta effect and the Seebeck effect.

The Volta effect is an open-circuit contact voltage (electromotive force) across a thermocouple due to the difference in energy levels in the two metals. Free electrons escape more easily from one metal than the other, which produces a thermoelectric voltage that depends on the two metals and is a function of temperature.

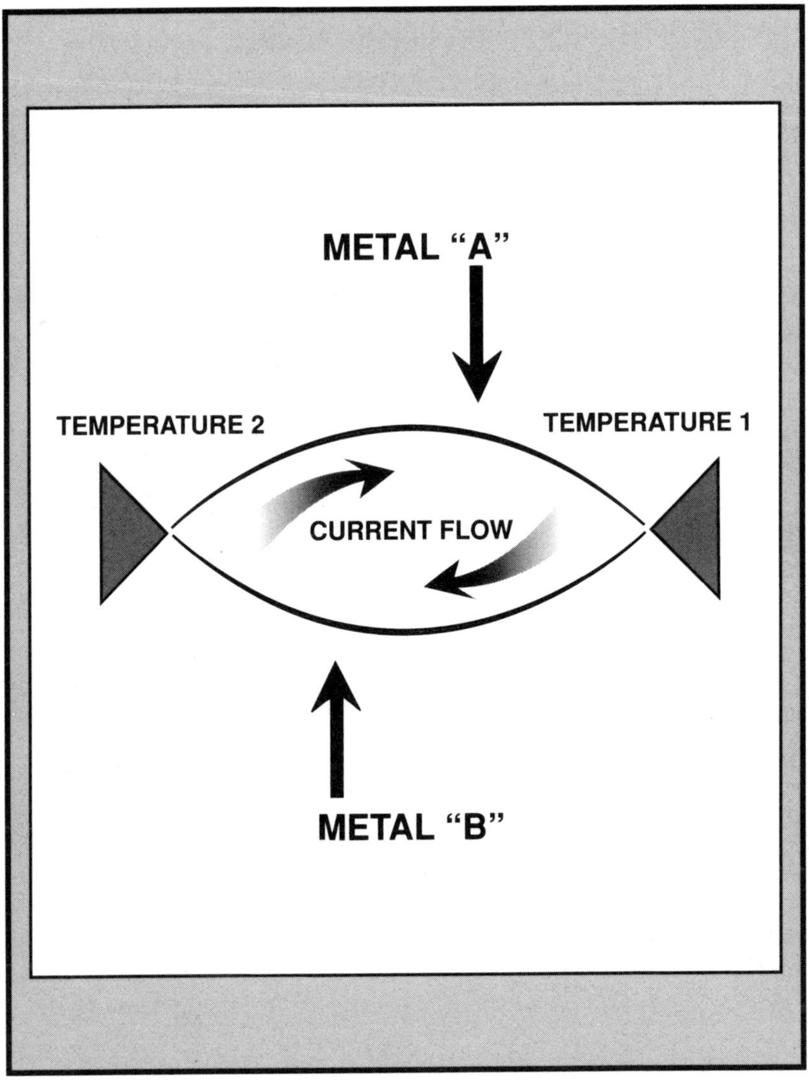

Figure 6-1. Schematic of a thermocouple

The Seebeck effect is the current that flows continuously in a circuit of two dissimilar metals when two junctions between the two metals are at different temperatures.

If one junction is held constant at a reference temperature or at a compensated reference voltage, the current that flows is a repeatable function of the temperature difference between the junctions.

One other effect is worthy of note: the Peltier effect. If current is forced by external means through a junction in the same direction as the current produced at the hot junction by the Seebeck effect, the Peltier effect will cool that junction; in the reverse direction, it will heat the junction. This effect is often used to compensate for variations in the ambient temperature of the reference junction.

Thermocouple output voltage tables for various combinations of metals are published by several standards groups; in the U.S. by the Instrument Society of America (ISA) and the American Society for Testing and Materials (ASTM). Thermocouple output voltage tables are generally relative to reference junctions at the ice point (0°C). Because of this practical convention, the reference junction is often called the cold junction, even though the reference could in fact be at a higher temperature than the measurement junction, for example, in cryogenic temperature measurements.

Thermocouple output functions are fairly linear over the useful ranges of the devices and can be compensated for accurately by either hardware or software in the measurement system. When reference junction temperatures other than 0°C are used (such as normal instrument ambient temperature), an additional correction is required via a compensation algorithm.

Thermocouple output voltages are small, ranging from about 7 to 75 microvolts per degree Centigrade. Figure 6-2 shows response of base thermocouples for common industrial thermocouple types, identified by the alphabetic designations at the ends of the curves.

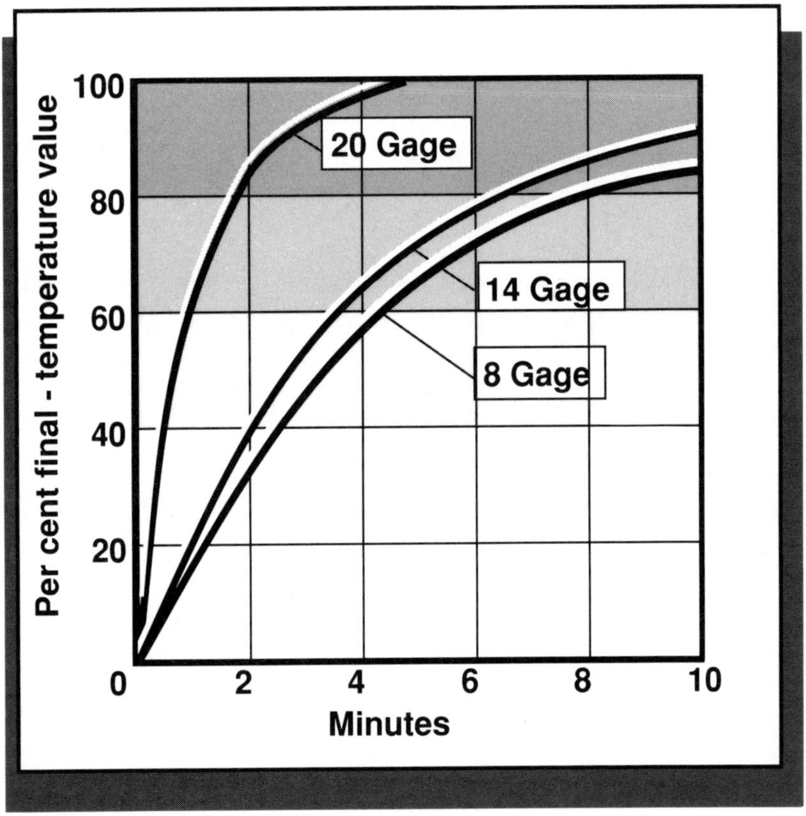

Figure 6-2. Thermocouple response times.

Each pair of dissimilar metals in contact comprises a thermocouple, including copper and solder. A practical measurement circuit must minimize the effects of such incidental thermocouples by assuring opposite couples in series at the same temperature. For example, an ordinary soldered junction of copper wires (if the wires happen to be separated by solder without direct copper-to-copper contact) comprises two canceling thermocouples.

A practical two-thermocouple measurement circuit using printed circuit boards and integrated circuits might well have some twenty incidental thermocouples in series, all of which must be canceling. This problem is not trivial.

If ambient temperature variations at the reference junction are small compared to the measured temperature, temperature compensation of the reference junction may not be necessary. For example, if the measurement temperature is 1,500C, a ±15°C ambient temperature variation at the reference junction represents only a ±1 percent error in the measured value.

For lower measured temperatures, or more precise measurements, there are two alternatives:

- Maintain the reference junction at a constant temperature. For example, in many field-mounted smart digital temperature transmitters, heaters and Peltier-effect cooling controls the ambient temperatures at the reference junction.
- Subtract a voltage equal to the change in the reference voltage due to the ambient temperature change. Where analog thermocouple wires are brought into the control room, this technique can be implemented by integrated circuits designed specifically for reference junction compensation. These circuits often include input/output isolation for safety.

Resistance temperature detectors (RTDs)

Resistance temperature detectors, or RTDs for short, are typically 500-ohm coils of fine platinum wired encased in a ceramic protective sheath. Like all metals, their resistance obeys a simple formula:

$$R = R_0(1 + \alpha_1 T + \alpha_2 T^2 + \alpha_3 T^3 + \cdots + \alpha_n T^n)$$

R = resistance at temperature T
R_0 = resistance at 0°C, and
$\alpha_1, \alpha_2, \cdots a_n$ are constants.

Two constants are usually sufficient to specify the resistance of platinum, while three or more are necessary for other metals.

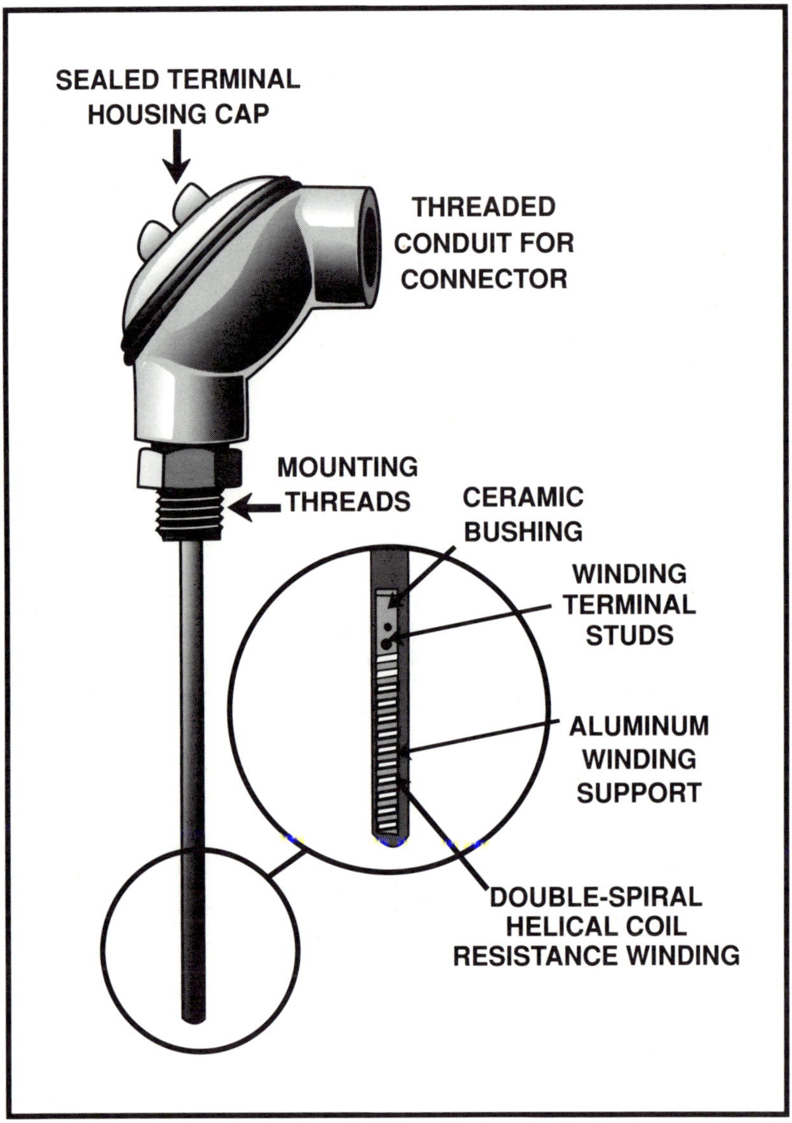

Figure 6-3. Typical RTD

The platinum RTD is limited to the range from -200°C to about 600°C, but over this range it provides the highest precision of temperature measurement. It has been used as the standard for the international temperature scale.

Thermistors

Thermistors are thermally sensitive resistors made from non-metallic composite materials with a high temperature coefficient of resistance, usually negative. The resistance is repeatable at a given temperature, and it changes exponentially as a function of temperature.

Most pure metals have coefficients between 0.3 and 0.6 percent per °C, while thermistor coefficients range between -3 and -5 percent in the same temperature range.

The thermistor-resistance temperature function generally is of the form:

$$R = \alpha R_0 \exp^{\beta/T}$$

where α and β are constants.

Thermistors are much more sensitive to temperature changes than thermocouples or RTDs, but they are also less stable over long periods of time.

Semiconductor temperature sensors

Semiconductor temperature sensors are made from bulk silicon, which has a positive temperature coefficient of about 0.7 percent per degree C, linear within 0.5 percent from -65°C to 200°C. Silicon junction diodes, or transistors with base and collector connected, exhibit a negative coefficient of resistance in their forward voltage drop. Again, this is linear with temperature over the range from -40°C to 150°C.

Since their physical mass can be very small, semiconductor sensors can detect very fast changes in temperature.

Radiation and optical pyrometers

Radiation and optical pyrometers are sensitive to heat radiation, as distinct from conduction or convection. Optical pyrometers are radiation pyrometers that work in the wavelength range of visible light, rather than the longer, lower-temperature invisible heat range of infrared wavelengths. In general, they are devices sensitive to visible light.

In the infrared range, any of the previously described temperature sensors can be used as pyrometers by insulating them from conduction and convection heat sources. Resistance elements like this (arranged to be sensitive to heat radiation) are also known as bolometers.

Unlike the temperature sensors previously described, radiation pyrometers are noncontact sensors, which makes them ideal in certain applications. For example, they can be made very small and arranged as arrays of semiconductor diodes, to conduct very fast and accurate scans of temperatures across a strip of moving steel in a hot rolling mill. Two-dimensional arrays of such diodes scanned with TV techniques provide infrared temperature images of process vessels and piping, with adequate temperature accuracy for safety surveys. Such instruments today can produce color thermal images as well as data for computer analysis.

A special problem in optical pyrometers is sensitivity to ambient or background temperatures, which can mask the specific temperature to be measured. One solution, known as dual-frequency optical pyrometry, involves modulation of a carrier frequency with the temperature being sensed, and modulation of a second carrier frequency with a reference, or background, temperature. Comparison of the measured signal with the reference signal then permits easy determination of the temperature to be measured. The technique can also be used in reverse, so to speak, to separate scintillation "noise" from slower changing process temperatures.

Pressure sensors

Traditional devices

Traditional mechanical pressure gauges based on deformable coiled "bourdon tubes" with mechanical linkages to indicating dials are still found on many process vessels. They are reliable, and they provide inexpensive pressure indication—as long as someone is present to read them. Similarly, liquid manometers still indicate differential pressures on many a laboratory bench.

Electrical pressure transmitters

The most common sensing elements in industrial pressure transducers and transmitters today are electrical because their measurements can be sent to controllers and computers some distance away from the point of measurement.

The primary sensing elements are strain gauges and capacitance cells, generally using metal diaphragms or bellows displaced by the liquid or gas whose pressure is to be measured. The diaphragm material must be resistant to any corrosive effects of the process fluid. Dual diaphragms filled with inert oils are often used to further isolate the measurement diaphragm from the corrosive process medium.

High pressure diaphragms may look like small drumheads, relatively thick compared to the diameter. At lower pressures, diaphragms grow larger in diameter and are corrugated radially to provide sufficient displacement for capacitance or strain gauge measurements. At still lower pressures, bellows designs allow even greater displacement sensitivity. A "push-pull" type of axially symmetrical diaphragm or bellows design is often used to provide canceling spring nonlinearities.

Smart pressure transmitters

Other pressure sensors, especially those used in smart digital

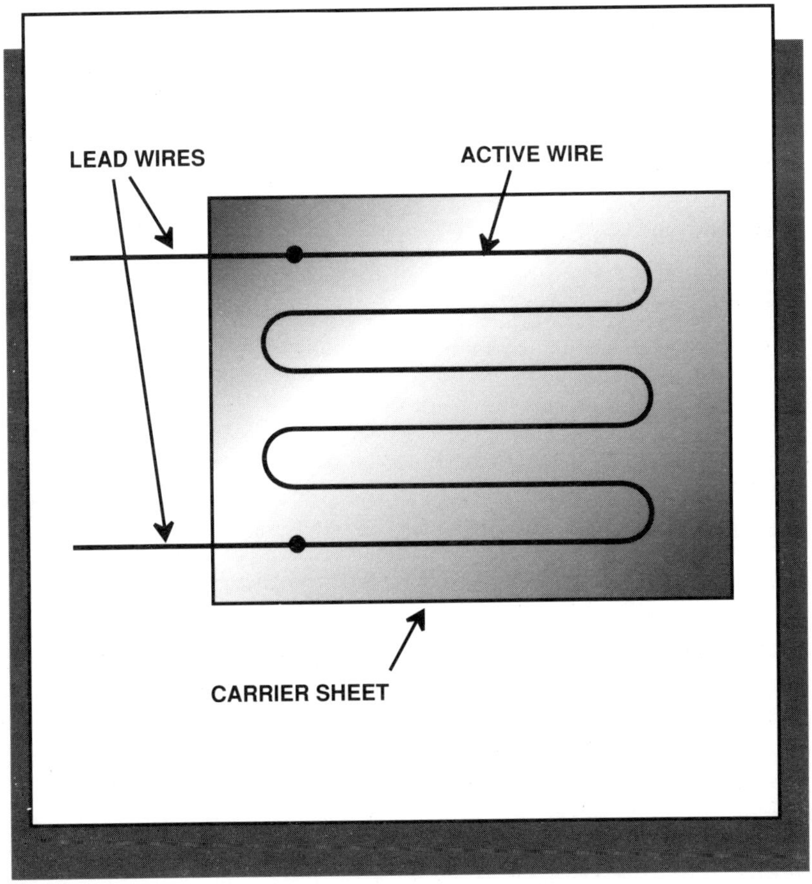

Figure 6-4. Strain gage schematic

pressure transmitters, use semiconductor strain gauges in two forms:

- Silicon strain gauge transmitters are built into silicon diaphragms using integrated-circuit manufacturing techniques. Temperature sensors and static pressure compensators are deposited directly onto the silicon wafer, together with an etched-out diaphragm and an analog amplifier. The output from the sensor amplifier is compensated for digitally by a microprocessor in the smart transmitter, using tables of values stored during manufacture. The final output is transmitted as either a 4-20 mA analog signal or as a digitally coded signal.
- Vibrating wires or silicon beams are stretched by attachment to diaphragms so that their vibration frequencies vary with pressure changes. Sensors based on vibrating wires or beams vibrate with a frequency that is a function of pressure. Strictly speaking, frequency is a continuous analog function, but it is easily converted into digital numeric values by counting the number of cycles in a fixed time interval generated by the smart transmitter's high-frequency clock. (By comparison, conversion of continuous analog voltages to digital numeric values is more complex, introducing other nonlinearities and uncertainties, whether for strain gauges, capacitance cells, or thermocouples.)

  In the near future, we expect to see digital transmission of all process variable signals become the standard measurement technique.

Level sensors

Continuous indication and control of the level of liquid standing in a tank or other process vessel has progressed a long way from dipsticks and gauge glasses, although even the newest electronic methods may remind you of these time-honored methods in the way they operate.

Level switches

Level switches, which open or close when the liquid level reaches the level of the switch, are both the simplest level-sensing devices and the basis of very sophisticated sensors. Level switches can provide on-off control of tank filling by directly operating solenoid or motor-actuated valves, or they can simply operate annunciators or notify a computer that a preset level has been reached.

Level switches can also be operated directly by the pressure head measured at the bottom of the vessel; by the buoyancy force of an immersed displacement body; or by the motion of a float. Each of these converts the liquid level to pressure or motion.

Pressure is measured by the motion of diaphragms or bellows as discussed above, and the motion of a float can be detected in much the same way. For single-point level indication, the diaphragm or float linkage can actuate an electrical limit switch. This switch can be monitored by a computer, announced by lights or horns, or it can open or close electrically operated valves, as mentioned earlier.

Another way to measure the level of nonflammable electrically conductive liquids is to detect the change in resistance between two electrodes when they are shorted by the liquid. Such devices are very inexpensive where applicable.

Level monitoring is a simple extension of the use of level switches. A series of level switches mounted at successively higher levels in the wall of a tank gives a relatively coarse measurement of liquid level from empty to full.

Electrical capacitance sensors

Many ingenious level indication techniques have been devised and marketed over the years. Electrical capacitance sensors are based on the principle that dielectric (nonconductive) liquids, as most organic and petrochemical liquids are, cause a change in the capacitance between two conductors as rising liquid level displaces the air above the liquid.

Level sensors using this technique measure the capacitance between small-diameter concentric sensor tubes extending from the top to bottom of the tank, with the outer tube perforated to allow immediate accommodation of changing level.

Capacitance is measured at a high frequency, for sensitivity to small changes, and at low electrical energy, for inherent safety against explosion hazard. The output of such sensors is continuous and linear with level, and measurement precision is limited only by the sensitivity of the electronic circuits.

Displacement sensors

Displacement sensors, mentioned earlier, use a long, thin, partially-submerged displacer body of constant cross-section, mounted vertically in the vessel, which produces a buoyancy force that increases continuously and linearly with immersion depth as the liquid rises in the vessel. The buoyancy force is measured by motion of the displacer against a calibrated spring, and can be converted to electrical signals by strain gauge or capacitance devices, as described above for pressure sensors.

Displacer-type level sensors move the electrical signals away from the process liquid. This improves the potential safety from electrical hazards compared to capacitance sensors, but it adds the nonlinearities of the motion-detection linkages.

Differential pressure sensors

Differential pressure sensors mounted both at the bottom of a tank or vessel and in the air or gas above the liquid, can provide an indirect measure of the depth of liquid if its density is known. Since the variation of density with temperature is known for any given liquid, accurate level measurement can be calculated if temperature is also measured. Add in a table of values for the shape of the tank at different liquid levels and temperatures, and you have a good way to calculate the total mass of liquid contained in the tank.

Computer-based hydrostatic tank-gauging systems are based on these principles. When this measurement process takes place in real time, as storage tanks are filled or emptied, it provides an accurate measurement of mass.

Liquid-to-liquid interface sensors

Liquid-to-liquid interface sensors are often required to measure the level of immiscible contaminants or condensates, such as the layer of water that accumulates at the bottom of oil or gasoline storage tanks. The mass of this water must be subtracted to determine the true mass and value of the product in the tank.

One high-technology technique to detect the interface between different liquids is to project sound waves through the liquids—sonar. The interface causes a change in diffraction of the sound waves, and hence a reflection or change in direction of the wavefront. This change can be detected with proper mounting of acoustic transducers, which makes it possible to measure the level of the interface continuously.

Thermal level sensors

Thermal level sensors come up naturally in the context of sensing the interface of two liquids or liquid and gas. Thermocouples are of little use, because there is generally no temperature difference across such an interface.

The thermistor is useful, however, since it is heated by the current necessary to measure its resistance. This heat is conducted away from the thermistor by the fluid the thermistor contacts, and the rate of conduction varies from air to liquid and from one liquid to another. As heat conduction changes, so does the temperature and hence the resistance of the thermistor.

While thermistors are inexpensive compared to sonar systems, they can only detect liquid interface changes at the level they are mounted. Each thermistor is only a single-point level detector—an electronic level switch.

Flow sensors

The rate of flow of fluids through pipes can be measured by two different standards, volume and mass. Both are in common use.

Volume flow measurements (the most common form of flow measurement) state flow rates in units per time period, useful in product cost determinations. Mass flow measurements state the rate in units per time period, compensated to standard conditions of temperature (and pressure, for gases). Mass flow rate measurements can be used to determine energy content of a gas, for example.

Differential pressure sensors

The most commonly used industrial flow sensor is the differential pressure sensor. It generally employs sharp-edged orifice plates inserted into the pipe and flow stream by means of flanges. The orifices have 90-degree corners at the inlet face and are usually relieved at 45 degrees on the downstream face.

In general, the flow rate through such a sharp-edged orifice is directly proportional to the square root of the drop in fluid pressure measured across the orifice. The square root characteristic must be linearized, as we described at the beginning of this chapter.

Pitot tube sensors

Pitot tube sensors, commonly used as air-speed sensors in aircraft, can also be used to measure fluid flow rates in pipes, but they are prone to clogging by build-up of contaminant materials in the flow stream. Pitot tubes share the square root flow rate characteristic of orifice plates.

Magnetic flowmeters

Magnetic flowmeters measure the flow rate of conductive fluids by impressing a magnetic field through the pipe. Flow produces a voltage at right angles to the magnetic field and proportional to the flow rate, which can be measured by electrodes in the walls of the meter.

The operation of magnetic flowmeters is based on Faraday's law, the same principle applied in electric motors. The output representing flow rate is proportional to the magnetic flux density, the pipe diameter, and the fluid flow rate.

An important advantage of the magnetic flowmeter is that is adds no internal obstruction to flow.

Early magnetic flowmeters used permanent magnets, which establish a fixed, unvarying magnetic field, but the measurement signal was small and swamped by noise from random electrical voltages induced in the fluid. Modern magnetic flowmeters use pulsed DC magnetic fields of alternating polarity. The device samples the signal during steady states, and carries out automatic zero-flow compensation with each cycle.

Vortex shedding flowmeters

Vortex shedding flowmeters insert a "bluff body" through the pipe into the flowing stream, sometimes circular or triangular with a flat side facing the flow. Such a bluff body causes vortices, or eddies, to break off on either side and downstream, which leave opposite edges of the body alternately at a frequency that varies directly with the velocity of the stream and inversely as the diameter of the body.

The vortices can be sensed as the differential pressure across the bluff body by means of capacitance sensors built into the body itself. The differential pressure signals are transformed into a pulse frequency and measured by comparison with a digitally generated time scale.

Vortex flowmeters have been gaining acceptance in recent years because they do not require that the fluid be electrically conductive.

Ultrasonic flowmeters

Ultrasonic flowmeters direct a beam of ultrasonic sound pressure across tipe, at an angle, to a receiving pressure transducer. The frequency of the ultrasound received changes according to the Doppler effect, in the same way the whistle of a receding train falls in pitch. This change in ultrasound frequency is proportional to the flow rate of the fluid.

This measurement is complicated by the velocity profile of the flow, which is fastest at the center of the pipe and slowest at the pipe walls. For any given pipe diameter, this effect can be compensated.

The ultrasonic flowmeter works best when there are entrained particles or bubbles in the flowing stream. No restrictions intrude inside the pipe, and strap-on models are available that make installation easy.

Coriolis-effect flowmeters

Coriolis-effect flowmeters measure mass flow of liquids, rather than volume flow. The Coriolis force is a gyroscopic precession-like force that arises at right angles to the direction of motion in a spinning inertial frame. For a simple example, watch how the water swirls as it goes down the bath-tub drain.

Coriolis flowmeters bend a section of the pipe into a circle and vibrate it about the axis of the straight section. A rotational torque is then caused by the Coriolis effect at right angles to both the straight section and the plane of the circular section. This torque is proportional (with the vibrational sinusoid subtracted) to the volume flow times the density of the fluid: that is, to mass flow. Density always depends on temperature, of course.

While electronically complex, Coriolis flowmeters are mechanically quite simple and place no restrictions in the flow path.

Figure 6-5. Mass flowmeter

Thermal mass flowmeter

A thermal mass flowmeter, or hot-wire anemometer, is merely a resistance wire of low mass which is heated by a constant voltage source. The resistance of the wire is constantly measured by the electrical current flowing in the wire and is proportional to the mass flow of the fluid in which the wire is immersed.

Response time of thermal mass flowmeters can be improved by servoing the current to maintain constant resistance at the wire.

Volumetric flowmeters

Volumetric flowmeters are positive displacement meters or pumps, such as residential gas meters, which typically make use of two bellows of known capacity, alternately filled with the gas and exhausted. For accurate measurements, some means of compensating for temperature must be included. Positive displacement pumps are also used commercially to deliver volumetrically measured amounts of products, such as gasoline to automobiles.

The objective of volumetric flow meters is totalized volume rather than flow rate.

Delivery of very large volumes of product from oil and gas storage tanks is generally done with mass flow rate measurements totalized over time rather than volumetric meters. Mass flow rate measurements are used because they can be more accurately compensated for temperature and pressure, and because the mass delivered, rather than the volume, more correctly represents the value that the customer purchases.

Fiber optics

Considerable research is taking place in the use of fiber-optics as a means of measuring temperature, pressure, and level.

The theory behind the research is that certain coatings change characteristics according to temperature or pressure or contact with certain liquids. For example, such a coating applied to the end of a fiber-optic cable can change the quality of light transmitted through the cable, and these changes can be measured and converted to engineering units. This technology offers great accuracy, plus stability over time. As the technology of producing such products advances, costs should decline to the point where they compare favorably with other measurement methods.

Control system considerations

The rest of this chapter looks at the other elements of automated systems: the pneumatic and electronic means by which data gets to the logic modules of the system; conversion and retrofit considerations when moving from pneumatic systems or PLCs to a DCS; the use of smart transmitters; modes of communication; and the operator interface to the system.

Pneumatic systems: old reliable

Pneumatic systems have been around many years running boilers on ships, operating wellhead chokes, and even supporting fairly comprehensive emergency shutdown (ESD) systems. A pneumatic system includes pneumatic sensors, pneumatic transmission tubing, pneumatic control panels, and, possibly, pneumatic to electronic interface switches. Basically, pneumatic systems operate from a compressed air circuit that incorporates the system logic. If part of the system fails, it must be identified and repaired or replaced wherever it's located, during which time the rest of the system may not be operational.

Because they use no electricity (and cannot spark), pneumatic systems have been used extensively in environments subject to combustion hazards. In practical terms, however, electronic systems can be built with explosion-proof housings and can be hermetically sealed to eliminate the possibility of combustion. Today, pneumatic systems are rarely incorporated in new designs.

Unlike electronic systems, the typical pneumatic subsystem has no failure detection features, making it suitable only for base regulatory functions where immediate notification of error conditions is not crucial. In simple pressure and flow regulation, for example, pneumatics are sometimes adequate: an operator adjusts the pressure and flow setpoints, and the system chugs along without further attention.

However, pneumatics falls short when automation requirements progress past the base regulatory level to advanced control and optimization. Pneumatic systems are simply not accurate, timely, or comprehensive enough to perform to the close tolerances demanded by optimized systems. Even the newer generations of pneumatic systems, which perform with greater accuracy, respond more quickly, and hold operating parameters better than the rugged systems of old, leave much to be desired: They are correspondingly more complex, more difficult to maintain, and have stringent requirements for source air. They offer better performance, but the additional complexity sacrifices some of the reliability that has been the hallmark of pneumatics.

Another disadvantage of pneumatic systems is their total life-cycle cost. A study on ESD systems reported in the March, 1989, issue of *Offshore* magazine indicates that an electronic ESD system would cost 46% less than an equivalent pneumatic system. Even making the electronic system redundant only increased its cost by 2-10%, still well under the pneumatic cost.

Redundancy is an important consideration. Since the cost of a particular pneumatic or electronic system is not a significant part of offshore facility costs (probably less than 1-2%), redundant systems may well be a very prudent expenditure. Redundant electronic systems are far less expensive than redundant pneumatic systems.

One other area where pneumatic systems fall short is their limited ability to be modified, expanded, or interfaced with computer applications. Modifying a pneumatic system typically requires shutting down the process while operations personnel get busy with wrenches and screwdrivers. Electronic systems can be modified, reconfigured, and enhanced from a keyboard, on-line and in real time, without even a pause to print out new documentation. And of course, electronic systems are designed to be interfaced with computerized applications.

Our conclusions here are obvious: electronic systems offer more flexibility, greater accuracy, and lower costs than their pneumatic counterparts.

Programmable logic controllers: quick fixes

Programmable logic controllers (PLCs) are relatively simple electronic devices designed to replace relays and timers. They are single-purpose microprocessors and offer a reliable quick fix for some isolated control problems. However, they rarely have the capability of carrying out sophisticated analog control functions using temperature, pressure, flow or similar measurements.

PLCs are typically programmed using "ladder" logic. Thus, ladder logic allows for certain actions that must be completed before others take place—using permissives and interlocks, for example. For some applications, PLCs also offer inexpensive solutions.

Early PLCs were environmentally hardened, but offered no redundancy and little fault tolerance. With the increasing capabilities of today's microprocessing chips, however, PLCs are expanding their capabilities to include high-level programming in Basic or "C," double and even triple redundancy, plus more complex regulatory functions and the capability to communicate with computer networks. However, these functions can be rather cumbersome compared to Distributed Control Systems.

In actual practice, most offshore operators do not see PLCs as an evolutionary step toward a Distributed Control System (DCS). They tend to see an either/or situation: They may select PLCs for sequential operations or independent shutdown systems, while they choose electronic DCS solutions for process control.

Figure 6-6. Programmable logic controller

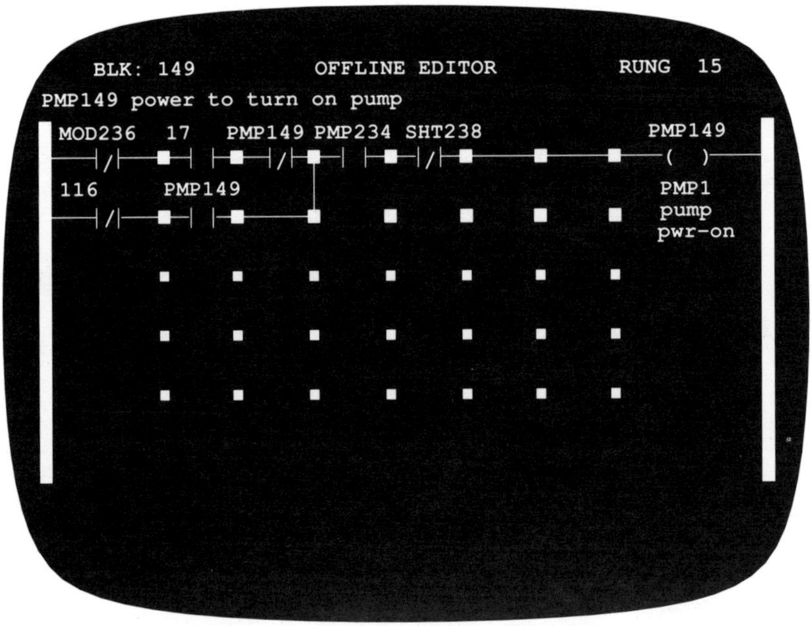

Figure 6-7. Diagram of PLC ladder logic

Using PLCs is relatively simple because each PLC usually relates to only one process. Also, since PLCs have been in existence for a long time, electricians and engineers are familiar and comfortable with them. PLCs also lend themselves to isolated and piecemeal planning where integration and overall facility management and optimization is not a concern. PLCs excel for high-speed control of single processes. They are ideal, for example, in high-speed numerical control and machining applications. But basic PLCs are inadequate in complex optimization and control situations. To construct a sophisticated system using PLCs, you need to add a computer with custom software.

One of the major problems in writing process-control code is creation of compatible network transmission protocols. In some cases, a host of PLCs each competing for message time can bring a system to its figurative knees.

The final disadvantage of PLCs in a large integrated environment is that PLC vendors in general do not take system responsibility. This issue can be extremely important in the decision to install and maintain mid-sized to large systems with complex interface requirements. The do-it-yourself, or hire a "systems integrator," approach of PLC suppliers can hamper integration, and it often leads to unforeseen additional expense.

Migration, conversion, and retrofit: whys and hows

Pneumatic systems are the oldest technology, PLCs are newer, and distributed control systems are the newest. Yet there is very little commonality in their operation and minimal compatibility in terms of interface.

Migrating from pneumatics to PLCs is a kind of lateral transition: You change the technology, but might gain little if anything in overall system integration. PLCs were originally designed to replace relay-logic systems. They should not be viewed as an interim step to a true distributed control system.

PLCs are often installed because they offer a degree of local or unit automation at a relatively low cost. However, when higher levels of integration are desired, the ongoing expense of connecting PLCs and maintaining communications to a central control facility can greatly exceed the cost of a well-designed DCS, even though the initial outlay for a DCS may be higher.

A quantum leap—the simplest path?

Those who lean towards the electronic DCS approach are the control engineers and instrument people who are most knowledgeable and who understand long range goals that may include integration on a broad scale.

It makes sense to start out looking at the long-range possibilities and planning for those from the start, even with small, local-control projects. Considering the modularity built into most new DCS systems, they are often the lowest-cost solution.

Surprisingly, perhaps the easiest migration path to new technologies is a quantum leap. A direct move from pneumatics to a DCS misses the grief of analog electronics and do-it-yourself computerization. In our experience, just about everyone quickly embraces new DCS technology with enthusiasm.

Conversion justification

Of course, the biggest difficulty in moving from old ways of doing things to a DCS may be justifying the cost. In a totally new installation, automation can be justified by the reduced size and cost of crew facilities, the reduced weight and structural requirements of the platform, the reduced need to generate electricity, savings on safety and liability insurance, and so on.

But in a retrofit situation, the old application is already in place, and installing a new system offshore and 100 miles or so from major land-based supply points is no trivial undertaking.

Many management and operations personnel also fear that installing a new control system means suspension of operations while construction is underway. They may believe that it will be more expedient and cost-effective to select an interim, pre-DCS solution. In fact, with proper planning and project management, modern DCS technology makes it just as easy, or easier, to reach the final solution directly.

Retrofits often make economic sense when the cost of finding spare parts and maintaining old systems have begun to skyrocket; where production is being lost because of the poor quality and excessive downtime of old systems; when safety considerations have sent insurance costs soaring; and when management decides that economics can be improved by upgrading control applications. Once the decision to retrofit is made, a DCS is by far the easiest system to expand for new applications and control strategies.

Managing the conversion process

Effective management of the conversion process starts by addressing the entire operation and how it has to be managed, and then developing a logical, sequential upgrade plan. It may not be possible to superimpose an effective advanced control strategy over an old basic control system. The pneumatics and PLCs may have to go. While it may be a difficult decision to throw out apparently functional equipment, the ultimate cost savings in reduced maintenance, more efficient management, and better information reporting will justify the investment.

You may be able to avoid replacing a well-maintained pneumatic or motor-driven control valve, but you will be less likely to avoid replacing old pneumatic pressure and temperature sensors. Once you replace part of a system you must then deal with interface problems and inherent actuation delays in the old equipment, which means that the best solution is usually a total retrofit. In virtually *every* case, you should replace the old pneumatic or PLC controllers, because they simply cannot be interfaced efficiently into the new totally integrated DCS system.

In all probability, you can use the same measuring and control points and locations, if not the actual devices. Use of the devices will depend on their maintainability, accuracy, and reliability. In some cases, you may not need any more sensors than your old system used. An advanced control strategy doesn't necessarily mean that you are measuring more variables, but rather that you are combining the information you already have to create a mathematical representation of the process. The idea is not to generate more information necessarily, but to bring all the information together in an orderly manner, so the system can analyze and manipulate it.

Of course, the planning process should provide for whatever new sensors are required to accommodate new control strategies and data requirements.

Conversions usually do not require disconnecting the system or suspending the process flow. On a pressure tap, for example, an isolation valve leads to a tube with the sensor. By turning off the valve briefly and installing a T in place of the tube, you can install the new electronic sensor in parallel with the old system, until everything is ready for a hot changeover.

In many industries and in many applications, new Distributed Control Systems have been installed and commissioned without ever losing production. Today's DCS manufacturers have designed their systems for rapid and efficient changeover.

The benefits of smart transmitters

In any large control system, the design of the overall communication system plays a crucial role in performance and effectiveness. There are two broad areas to consider: the type of data to be transmitted, and the communication method used to transmit it. We'll talk first about data, which might appear at first glance to be a simple problem. In fact, there are many questions to answer. Consider, for example, a thermocouple measuring temperature. Do you transmit the temperature as a pneumatic signal or an analog electrical value directly from the thermocouple? Or do you perform basic range, scale and conversion functions at the sensor location and transmit a digital value for the temperature? Or do you go one step farther and track changes in temperature, so you can transmit values only when they change?

In the past, computational and safety limitations at the sensor location meant that the data transmitted was basically raw data in analog form. Today, smart transmitters make it possible carry out detailed data analysis right at the sensor location.

Smart transmitters are available to measure a wide range of process variables, including pressure, differential pressure, temperature, mass flow, vortex flow, pH, and conductivity.

Smart transmitters vs pneumatic and electronic analog sensors

Smart transmitters are superior to both pneumatic sensors and electronic analog sensors in fundamental ways:

- They offer far greater accuracy and reliability, due to their built-in analog-to-digital conversion capability.
- They are more stable and perform without drift.
- They can share a digitally-based communication line.

In addition, smart transmitters offer enough other advantages that they justify the replacement of older devices fairly easily.

Functionality for complex tasks

Because they have considerable built-in microprocessing capability, smart transmitters can do things that can't be done any other way:

- Smart transmitters handle complex computational functions, such as signal conditioning and linearization, for which ancillary control devices were once required. This simplifies operations and means that less data is transmitted, which keeps the communication network from bogging down.
- Ranges can be changed depending on process conditions. If you are running at 50% capacity, you can change the range of a transmitter to get a more accurate measurement. In fact, digital control systems can do that automatically without operator intervention.
- Smart transmitters can even diagnose their own problems. In new digital technology, the device can use internal redundancy to pinpoint problems down to the actual part that needs replacing.
- It's easy to change control strategies. Instead of resetting devices or installing new ones, most changes can be handled from the control room.

Remote operation

Other advantages of smart transmitters stem from the fact that they can be configured and maintained from a distance:

- Configuration and calibration from the central control facility frees the user from costly maintenance chores at remote locations.
- It takes less time to calibrate smart transmitters. When you calibrate an analog device, you're out on the platform with a screwdriver. With a smart transmitter, you type a command

into the computer. If you have a lot to do, this is a significant time savings.

- Smart transmitters can monitor their own functions and transmit diagnostic information to the central control facility, which makes possible predictive maintenance systems that can dramatically reduce total maintenance costs.
- Maintenance is further simplified because the configuration of each smart transmitter is maintained by the central system. After a maintenance person plugs in a new part, the control operator simply downloads the configuration to get everything running again.

Overall system enhancement

Smart transmitters not only work better as sensors, they allow the entire system to function more effectively:

- The computers and other data processors in the system maintain their capacity to handle critical monitoring and control applications, because they are not burdened with vast quantities of data scaling, conversions, etc.
- Smart transmitters can be multiplexed, which saves wiring costs. For example, 15 different smart transmitters may share one twisted pair: If those 15 sensors were analog, they would each require a separate twisted pair.

As we will see in the following section, the use of smart transmitters can dramatically improve the performance of the system's communications network.

The communications network

Choosing a communication network to meet process management requirements is a perplexing problem: The network must meet requirements for effective process control while retaining enough spare capacity to handle the worst-case plant upset. It must accommodate many levels of data from simple point-to-point data transfers to fast alarming for critical situations.

Whole books have been written about this, but we will limit our discussion here to lists of options in key areas: network services, media type, network type, redundancy, and data compression.

Network services

A network can provide a variety of services to users, from local communication to a process unit to microwave and satellite links with distant parts of the world. Basic network services include:

- Data transport, to ensure accurate reporting of process variables and control signals.
- Programmable device messaging, to ensure that only the correct device is receiving or sending data.
- File transfer that maximizes the amount of data that can be transmitted at one time and which permits communication with foreign devices.
- Resource sharing, allocating the use of computing and processing resources to provide the fastest and most efficient control function.
- A multiple access distributed database to minimize the access time for data that must be shared among process and control units.
- Electronic mail, to allow non-process control messages to be transmitted at a lower priority within the same process network. This avoids expenditures for multiple data pathways.

Media type

There are many ways to build a network, from twisted pair cables to fiber optics. The following table summarizes the options.

Media type	Range (in meters)	Bandwidth	High freq. loss	Noise immunity	Relative cost	Comments
Twisted pair Shielded	<100 m <1000 m	<10 MHz <100 MHz	High	Very good	Low	Inexpensive for short range
Twinex	<1 km	<200 MHZ	<50-100dB per km	Very good	High	Can be compatible with IEEE 802
Flat multi-conductor	<100 m	<10MHz	High	Poor	High	Can transmit parallel bytes (IEEE 488)
Coaxial	<2 km <1 km	<400 MHz <1 MHz	6.5 dB/km 60 dB/km	Fair	High Low	Can be compatible with IEEE 802
Fiber optics	2-10 km	>500MHz/km	<8 dB/km	Excellent	High	Costs are dropping with new technology
Telephone	—	<50 kHz	High	Poor	Low	Low cost

Figure 6-8. Communication options

Network type

Network type refers to the manner in which messages are prioritized and delivered. Network types include the following:

- Serial highways treat all data equally (without priority) and move data as it comes in. This is the simplest way to transmit data, but it is useful only for very limited applications; for example, to connect a single loop controller to a console or workstation. Expansion capability is limited and interdevice communication rates are typically slow, because there is no way to cluster data for more efficient transmission or to prioritize individual messages.

- Traffic controlled networks are governed by a traffic director that sits at the hub of a "star" configuration to reduce the number of highway interface points. Each point of the star may be a single loop (serial highway), or even a multiple loop with its own (sub)traffic controller. In the latter case, the system becomes a hierarchy of traffic controllers.

The disadvantage of this system is that data transfer is limited to the time allotted to each point. If communication is not complete and time elapses, this point must go into a wait state. This can be a real problem in the event of a process upset where large amounts of data must be transmitted.

- Multidrop highways allow multiple traffic controllers to connect to the same highway. In this case, there is no star configuration—each traffic controller becomes the master for a predetermined period. Control is granted to each controller in turn by passing "tokens" related to the priority (time or importance) of the messages to be transmitted. This configuration has the advantage that all nodes can send and receive directly to and from all other nodes, but it is subject to seizing up with "token choke" if a major alarm situation occurs.
- Multidrop ring highways. In this type of configuration, all nodes are connected in a continuous loop. Messages are sent by a node and retransmitted at each successive node along the ring. Data security is easily checked, and the ring acknowledges all messages, ensuring that only the correct data is acted upon.

Ring networks achieve higher data transmission rates than multidrop highways, and heavy loading does not cause appreciable delays. The apparent data throughput of this type of structure is at least 10 times faster than the others described above. When configured in a redundant manner, this structure is the fastest, most reliable communication scheme available.

Redundancy

Redundancy is an important consideration in network design. Systems using a multidrop ring generally include dual redundant communication lines as standard, while token-passing multidrop networks typically rely on high Mean Time Between Failures (MTBF) and do not provide redundant lines.

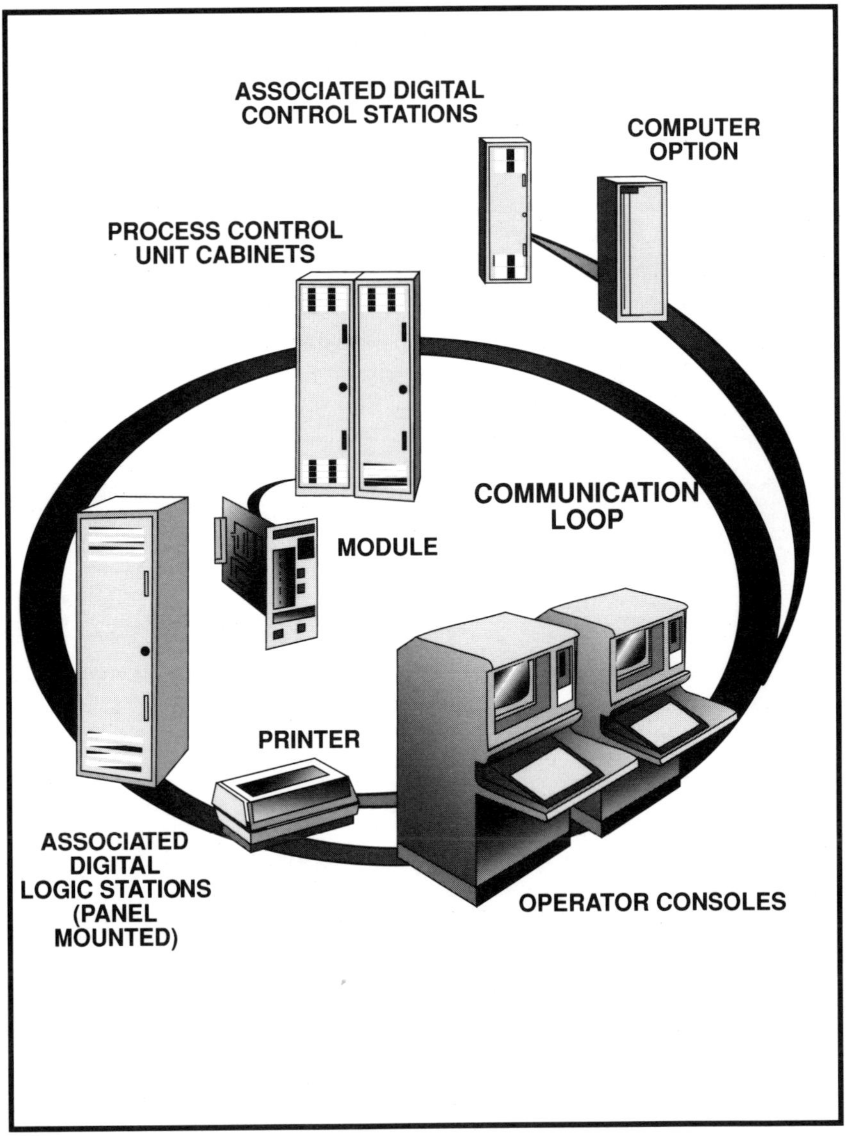

Figure 6-9. Communications network

In the dual redundant mode, data is transmitted on two highways simultaneously. The receiving node decodes and compares the data from both channels. IF both channels match, the data is accepted as good and used.

If the channels differ, the receiving node evaluates error-checking codes to determine which channel is correct. It then discards the data from the incorrect channel, sends out an alarm message, and retransmits the good data. In this way, errors are not propagated throughout the system.

In addition, as each incoming message is received, it is acknowledged and retransmitted to the originating node, which then removes it from the communications loop. This "deterministic" acknowledgement and removal is unique to certain protocols. It is the most secure method of communication, and highly desirable for process control applications.

Data efficiency enhancements

The way to optimize a network is through the use of data compression techniques. There are several approaches:

- Exception reporting—Data is transmitted only when it has changed significantly from a previous value. Smart transmitters manage control actions at the local level, reporting only the deviations from preset limits to the communications loop.
- Packetizing—Here multiple packets of information are combined into a single transmission message, and transmitted to multiple locations. Packetizing reduces the total highway loading, because it reduces the traffic control and error-checking activities that take place for each message that goes out on the communication loop.
- Prioritization—All types of data can be assigned priorities based on time value, location, emergency status or other measurement of importance. For example, alarms have highest priority, followed by significant unplanned changes in key process variables, and so on. By handling data based

on priority, the system can ensure that the most important information is acted upon first.

As you can see, the structure of a communication system is a complicated issue that is critical to the effectiveness of real-time process control. It is important to select a DCS system with an integrated communication design that can accommodate both small and large applications.

The operator interface

If there is one single factor that creates acceptance or rejection of any control system, it is the operator interface. Regardless of the quality, specifications, or performance of the control system, the visual presentation of information to operators and other users is *the* most critical factor in ensuring the success of the installation.

There are many ways operators can interact with process control systems. Simple hand-held units can be plugged into local processing units throughout the platform to check and recalibrate sensors. Panel-mounted visual-display units can provide continuing display of process variables and also provide a means of calibration.

In the control rooms (local or remote), operator interfaces can range from standard PC QWERTY keyboards with color graphic displays, to complex interactive consoles including high-resolution CRT graphic displays of process diagrams and pertinent information for 30,000 process variables.

Many of these operator consoles use specially-designed keyboards so that operators can simply push the right button instead of worrying about typing in a command. Single push-button keystrokes can implement control actions and access critical system functions such as alarm management and system status displays. Pull-down menus, touch-screens, and joysticks also simplify operator actions.

Some systems use a hardwired type of graphic display in addition to CRT displays. One recently designed safety system for an LNG terminal uses a hardwired graphic with a maze of

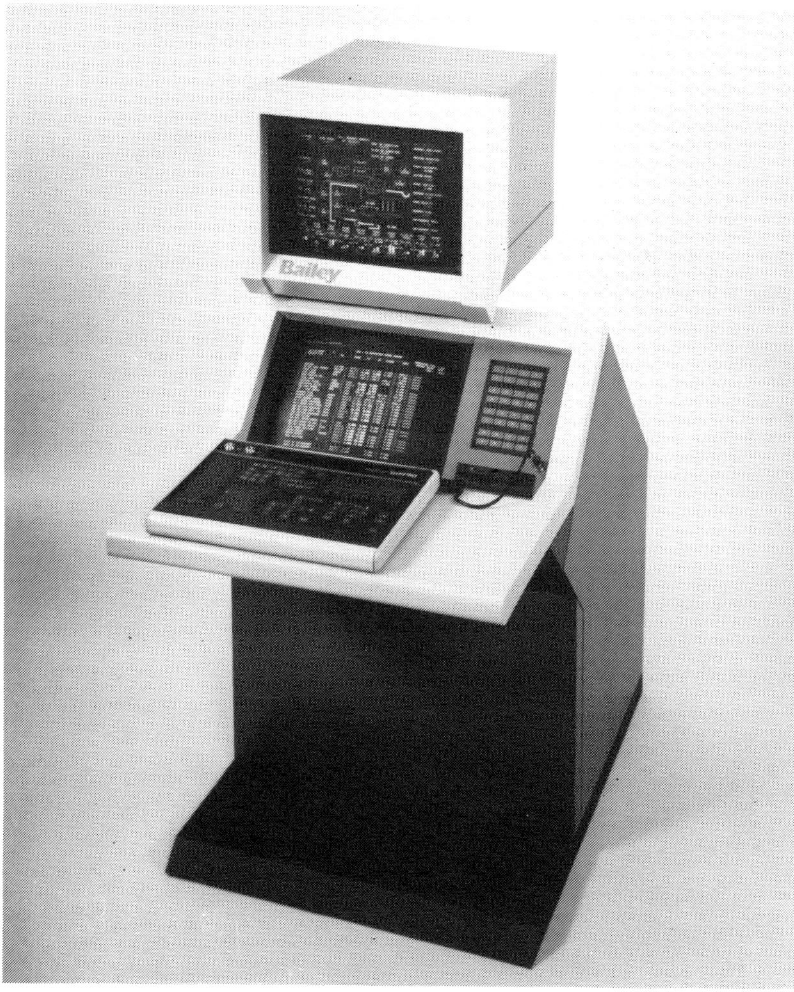

Figure 6-10. Operator interface unit

red lights built into an overall diagram of the plant to show the presence of gas. By looking at the display of lights, operators can see and track the movement of a gas cloud that may pose a hazard to surrounding areas. To "zoom in" on an alarm, a CRT shows a detailed diagram of process and alarm conditions in a particular location. Another CRT gives instructions to the operator for closing down a particular subsystem or turning on extinguishant systems, etc. Visual input is provided by closed-circuit TV, and audio information on the plant PA system (via the computer's speech synthesis program) completes the alarm system.

With the advent of computerized "windowing" visualization technology, operator consoles can provide a window to *any* networked database, which allows for unprecedented integration of process management applications, wherever they are in the system.

In the future, "heads-up" displays like those used in airplane flight controls may provide even more sophisticated control options. And holograms may provide three-dimensional visualizations of platforms and processes for control and troubleshooting.

Summary

In this chapter, we looked at the wide variety of sensing devices used in process control systems. We then discussed briefly two older methods of using and transporting the information gathered by sensing devices—pneumatic systems and programmable logic controllers—focusing on the shortcomings of these systems in today's environment.

The discussion of converting to a new DCS-based approach considered alternate migration paths, with a strong emphasis on the value of smart transmitters as the basic building block for any new system.

We concluded the chapter with a brief look at communication alternatives and the importance of the operator interface.

Having thus covered the first two levels of the structure we built in Chapter 4 (Base Regulatory and Supervisory Control), we will move on in Chapter 7 to process control considerations, and in Chapter 8, to optimization and facility management.

Chapter 7

PROCESS CONTROL CONSIDERATIONS

In the last chapter, we talked about the sensors on the platform, the communications network, and the advantages that distributed control systems offer over pneumatic and PLC implementations. In this chapter, we'll examine some of the things that affect all process control systems: robustness, redundancy and fault tolerance, alarm prioritization, and useability considerations.

Robustness

Robustness is a control system's resilience to change. In one sense, this relates to changes in process parameters; for example, if a time constant or process delay increases, the control system must still be able to maintain adequate control of the process.

A robust process control system also continues to provide adequate control even though components may fail, or while system testing, maintenance, or replacement of equipment is taking place. Ideally, uptime is as close to 100 percent as possible. When the system senses an alarm condition, it must be able to respond immediately by logging the alarm, notifying the operator, and, in some cases, initiating remedial procedures. In the case of fire, for example, the system should be designed to automatically turn on extinguisher systems without waiting for an operator command.

At the same time, the system should not be fooled by fault conditions within itself or by spurious signals from sensors. In other words, it should be "fault tolerant." It cannot cause nuisance alarms and shutdowns that only create additional work and an opportunity for operator error.

All of these attributes contribute to the "robustness" of a system. Robustness is an all inclusive term that describes the overall ability of a system to keep operating, regardless of what happens, whether it be a typhoon, an undetected electronic failure, a complex maintenance procedure, or an explosion at the wellhead. Robustness can be described in terms of reliability, availability, and maintainability.

Reliability

Webster defines reliability as "fit to be relied on." More formally, reliability relates to the probability that a system will perform satisfactorily for at least a stated period of time when used under stated conditions. For example, reliability specifications may call for less than one hour of system downtime per year over a period of 10 years in an environment where temperature ranges from 50°F to 90°F.

Reliability depends on the quality of the system's components and tight manufacturing control in assembling those components. To a certain extent, it must also be designed in. Some ways to improve reliability include:

- Using only high-quality components. This means that components are produced under a "quality process" that uses advanced control techniques like Statistical Process Control to ensure product consistency and performance.
- Special enclosures and component "hardening."
- "Derating" components, which means using them at a fraction of their rated capacities.
- Maintaining ambient temperatures at constant and reasonable levels, preferably close to 20°C (68°F).
- Keeping vibration and mechanical stress to a minimum.

Figure 7-1. Offshore platform

- Using duplicate or triplicate components for the same task. Redundancy approaches can range from duplicating certain critical components to duplicating the entire system. We will discuss redundancy in greater depth later in this chapter. Suffice to say, redundancy can be extended to whatever level required.

Availability

Availability goes beyond reliability to describe system uptime. It relates to the probability that the system is operating satisfactorily at any given instant of time when used under stated conditions. In calculating this probability, the total time considered generally includes both normal operating time and active repair time.

Some safety systems, for example, are built for 99.99% uptime—which is less than one hour down per year! Systems of this type often utilize components that test themselves and each other on an ongoing basis. Errors or marginal conditions are reported back to the central control room so the failing or marginal component can be replaced immediately.

Any time a system is undergoing routine maintenance functions, even self-diagnostics, it may be unavailable. To circumvent this problem, some systems are designed to abort self-testing if a process alarm is sensed.

A computer with low reliability can still be part of a highly available system if it is backed up by redundant computers. Of course, it's much better and easier to achieve good availability by starting with reliable components.

Maintainability

This is the third critical factor. If problems are found in a system, they must be easy and quick to fix, preferably without taking the system off-line.

Maintainability must be designed into a system. Control system elements must be built as modular, easily replaced units. Preferably, they will be modules that can be plugged in under power. This eliminates the problem of having to power-down the system just to replace a marginal component.

Fault tolerance

Fault tolerance is the ability of a system to keep doing its job in spite of faults within the system or spurious signals from sensors. A fault tolerant system is one that avoids nuisance shutdowns and creates a high level of confidence in users.

There are two types of faults: overt and covert.

Overt faults are easy to spot and correct, so they generally "fail to safety." For example, if a sensor quits sending data, an operator will notice it quickly and replace the sensor.

Covert faults are the subtle faults that "fail to danger." They are often difficult or impossible to spot. For example, if a temperature sensor continues to transmit a normal value while the actual temperature in a reactor is increasing out of control, we have a covert fault. Process control systems, and safety systems in particular, should be designed to eliminate the possibility of all faults, but especially covert faults.

Redundancy and voting systems

One way to reduce the chance of covert faults is through redundancy. Generally, anytime you have more than one device doing a job, you lessen the probability that failure of a device will take down the system.

There are several approaches to redundancy.

No redundancy

The simplest approach. You can get by having no redundancy in your system when failure of the process poses no particular hazards and downtime is easily tolerated. Within a single non-redundant system, reliability, availability and maintainability features can make a significant difference in overall uptime.

Standby redundancy

In this approach, a functioning element has an equivalent unit standing by. If errors are detected in the main unit, the standby can be put into operation quickly.

While this generally offers better overall uptime than a non-redundant approach, it has some significant drawbacks: The worst is that the redundant system isn't available to help diagnose failures in the primary system. That must depend on the main unit's essentially nonredundant failure detection system.

Another problem can occur if the standby unit is put into service before the problem on the main unit is diagnosed and fixed—

whatever caused the main unit to fail could also cause problems in the standby.

Hot-standby redundancy

In hot-standby redundancy, two systems (on-line and standby) are always in operation, running in parallel. The standby unit is ready for immediate operation when activated by an automatic switchover mechanism.

In some cases, only the hot (on-line) unit actually collects data and sends out control sequences. The on-line unit continually updates the standby unit, and the standby unit continually monitors the on-line unit. If the standby unit detects malfunctions in the on-line unit, it immediately takes over with full knowledge of the expected control philosophy and the current operational parameters.

In other cases, both the on-line unit and the standby unit receive data from the same sensors or from redundant sensors. This ensures that the standby unit is fully operational and up-to-date with current computtional values, without having to depend on the on-line unit for periodic updates. In this arrangement, it is possible that both on-line and standby units believe they are functioning properly but are producing different outputs. Some selection algorithm is generally used to decide which unit is incorrect and shut it down.

The hot-standby approach is frequently used in offshore operations because it provides "bumpless" transfer of control within micro-seconds. The failed unit may then be serviced, and it or another unit may be put into service as the new standby.

Active parallel redundancy

In active parallel redundancy, two or more components or channels all do the same thing at the same time.

A dual system can work by having each channel monitor both itself and the other. If a channel finds that it's not working, it shuts itself down and transfers control to the other. If it finds the other not working, it reports the fact and keeps going.

Dual systems can also use a voting scheme, either one-out-of-two voting or two-out-of-two voting:

- One-out-of-two voting means that if either of the channels detects an alarm, the system takes action. This offers the greatest protection against hazards, but it does not protect against nuisance shutdowns.
- Two-out-of-two voting means that both channels must agree before action is taken. This prevents nuisance shutdowns, but it may miss a hazard. In this scenario, the system should be continually testing itself so that spurious differences between two duplexed components can be repaired.

A three-channel or triple redundant system generally uses two-out-of-three voting to provide greater fault tolerance while greatly reducing the risk of missing a potential hazard. Systems like this are, of course, more expensive, and are used mostly for extremely high-risk applications, such as Emergency Shut-Down systems.

In some cases, even higher levels of redundancy are used. The U.S. National Aeronautics and Space Administration uses three-out-of-five voting to launch a space mission, and some nuclear installations use seven-out-of-nine voting.

Full redundancy

The redundancy types discussed above refer to redundancy in individual components of a system. "Full" redundancy means that all components in a system are redundant. Relatively few offshore automation systems employ full redundancy, if only because most sensors are not duplicated; redundancy here relies on having a number of sensors close enough to each other that fault conditions are not missed if one of them fails.

Digital DCS reliability and availability factors

With the advent of smaller, faster, and cheaper semiconductor technology, today's all-digital distributed control systems often employ a high level of internal redundancy. In many cases, this is carried right down to the level of the smart transmitter, so the operator in the control room can pinpoint failing components to the exact module where they occur, whether that be the sensor itself, data conversion electronics, or some processing function higher up the line.

Alarm prioritization

One of the ways process control systems differ is the sophistication of their alarm prioritization system. Many simple systems just send all alarms to the operator, who is expected to filter everything and handle the most important things first. Even mainframe systems often work this way because the computer spends most of its time in control and analysis functions, with little left over for alarm processing.

Distributed control systems have the advantage of multiple parallel processors, each of which manages its own segment of the process, including the relevant alarms. Even so, different DCSs apply varying degrees of attention to alarms.

One of the most useful ways of prioritizing alarms is time phasing. Often one alarm creates a situation that causes more alarms, which causes more alarms, and so on. You don't want to be inundated, you want to concentrate first on the one alarm that started the whole thing and how to fix it.

A well designed alarm interface can be overseen by a single operator. It will time-sequence and prioritize alarms, and it should also condense information by combining several variables into single, more meaningful variables. For example, it is more useful to know that a reactor is about to explode than it is to know that the temperature has increased by 50 degrees and the pressure by 150 psi.

Alarm systems should also be tied to automatic shutdown procedures if they are appropriate to the application. If the operator is unaware of a critical problem or unable to take action for some reason, the system should be able to do it.

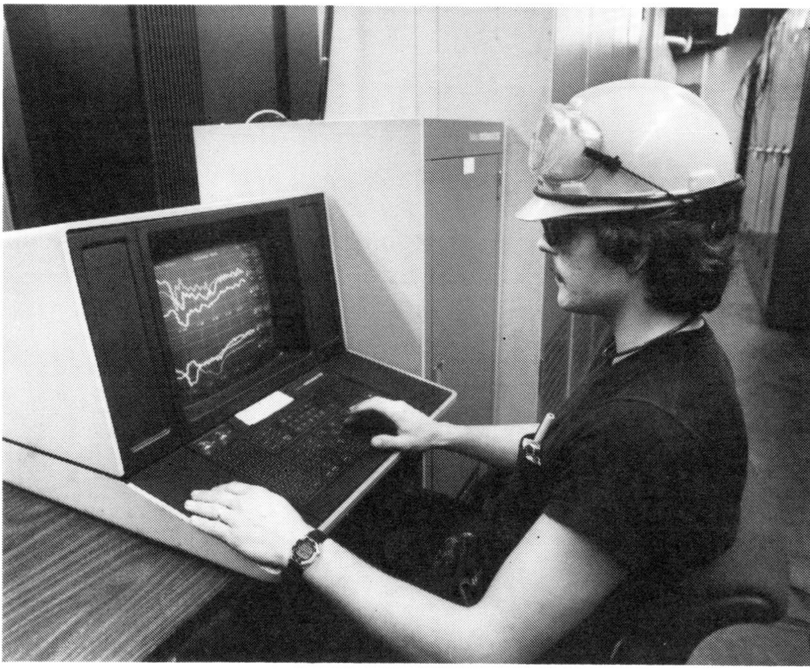

Figure 7-2. Operator console

Documentation and useability

Each system type has its own useability scheme; different methods of programming and documentation. We discussed mainframe programming briefly in Chapter 5, and the ladder logic of PLCs in Chapter 6. Distributed control systems usyet other methods.

Some DCSs on the market today still require an engineer to enter numbers and codes directly into the controller via a numeric keypad or with lines of code. The engineer will key in the input number, followed by a sequence of function codes for the function to be performed: square root, sum, unit conversion, etc., followed by an output code. For documentation, the system prints a list of the codes entered.

This system has several pitfalls. First, the engineer must go out into the operation where the controller is located in order to enter the control sequences, not always an easy environment in which to work. Second, there is nothing visual to guide the engineer through the process. Many of these systems do not support CRT screens for data entry. Third, the documentation produced (the printed lists) are difficult to read and check for accuracy: It's hard to mentally translate a list of numbers to an idea of the process taking place. And finally, the system can be very time consuming. If you have 1,000 temperature loops that differ only in minor respects, you must still enter each one individually.

Newer systems use PC-based engineering workstations for data entry, and it's difficult to overstate the advantages of this approach:

- A PC-based workstation with a CRT screen can operate using block logic that is second nature to most process engineers. By drawing a process and instrument diagram (P&ID) on the screen, engineers can create the process control strategy as they go. The workstation has pulldown menus for drawing (circles, rectangles, etc.); choosing icons to include (symbols for valves, pumps, compressors, etc.); executing mathematical routines (engineering conversions, square root, sum, etc.); and even ladder-logic modules.

- The block diagram on the screen is a strong visual aid that improves the accuracy of programming and dramatically reduces the amount of time required.

- Once the diagram on the screen is complete, the information can be downloaded to the process controller via floppy disk, or, in more advanced systems, directly from the engineer's workstation. This means that programming can be done in a suitable environment, not on the plant floor.

- System utilities support copying; for example, a temperature loop can be copied to allow entry of a different tag name and set point, so that the entire loop does not have to be recoded.

Figure 7-3. Distributed control system

- The system is relatively simple to learn, and, in many cases, clerical personnel can take over repetitive jobs (such as copying 1,000 temperature loops) to free the engineer for creative work.
- The system is self documenting. The final process diagram on the screen not only provides the instructions for the process control computer, it can be printed or plotted for "instant" documentation. Any time a process strategy is changed, new documentation can be produced immediately.
- The system is self-verifying. Each microcontroller can be queried for its program and automatically compared against the engineer's P&ID. Any differences can be printed for the engineer's review.
- Documentation and useability considerations are often overlooked by those choosing a system for the first time. It's easy

to be captivated by raw system speed or power or special features or low cost. But ultimately, a system may be only as good as it is easy to use.

Summary

In this chapter, we have discussed topics crucial to a good understanding of process control systems: Robustness (reliability, availability, maintainability), fault tolerance, redundancy and voting systems. And we have looked at characteristics of Digital Control Systems; in particular, their capabilities for alarm prioritization, documentation and general ease-of-use.

These are the basic concepts underlying any workable approach to offshore automation systems. Careful attention here can help ensure your success in process optimization and facility management applications.

Chapter 8

OPTIMIZATION AND FACILITY MANAGEMENT CONSIDERATIONS

Optimization and facility management occupy the third and fourth levels of the pyramid defined in Chapter 4. The two levels are closely related: Optimization helps gain the greatest advantage from a single process or group of processes, and facility management expands that concept to the entire offshore platform.

The two key tools for optimization and facility management are statistical process control (SPC) and artificial intelligence (expert systems).

A brief history of Statistical Process Control

Recent increases in global competition have fostered the spread of a management philosophy based on the idea of the "quality process" as developed by Walter Shewhart, W. Edwards Deming, Philip Crosby, Joseph Juran, and William Conway. While their philosophies are not always consistent, and their definitions of quality differ, these men have been instrumental in creating new approaches in both manufacturing and service industries.

Initial development of the quality process took place in the 1950s, not with the support of American captains of industry, but in Japan, a country willing (for obvious reasons) to try a new way of doing business. Their success has spurred the rest of the world to adopt similar processes.

Important among the tenets of the quality process is that quality means continual improvement. And one of the most important tools to achieve continual improvement is Statistical Process Control, or SPC.

The conventional way of making a product or running a process without SPC is to just do it; then check later to make sure the product meets specifications. This approach is often referred to as "inspecting quality in," as opposed to building it in. Bad batches are thrown out, recycled, or sold as seconds. It's an expensive way of doing business, and incredibly ineffective in the face of competition—to which the American automotive and electronics industries can attest.

Traditional (off-line) SPC offers powerful retrospective analysis tools that can help find and eliminate problems that cause products to go "off-spec." New on-line SPC systems, offered with today's most advanced process control systems, make it possible to identify and rectify problems while they are happening, and before they can have a substantial impact on process economics.

We will first discuss the general principles, charts, and reports of SPC. Then we'll describe how the system can be adapted to on-line use in a DCS environment, and the advantages that it offers.

Fundamental principles of SPC

SPC is based on the principle that there is a usual, quantifiable level of variation in every process parameter. This normal variation includes both random variation and variation due to special causes.

- Random variation is inherent in a process; for example, the temperature of a reactor may fluctuate normally by 5 degrees.
- Variations beyond normal random variations are due to specific causes—a stuck valve or a malfunctioning motor, for example.

Quantifying normal variation offers a way to manage a process to produce your product most profitably—either by eliminating special causes or by modifying the process itself to reduce the effects of special causes.

If the normal variation of a critical product quality parameter is greater than the tolerance of its manufacturing specification, the product will be very expensive to make, because you'll spend a lot of money testing it and throwing out or reprocessing off-spec material. On the other hand, if the normal variation is substantially less than the manufacturing specification tolerance, you may also be wasting money controlling the process too tightly.

SPC offers solid statistical tools to quantify and manage this dilemma. Control charts, SPC alarms and Pareto analysis can all be used in the on-line environment. Cause and effect correlation studies and process capability studies are additional tools for off-line analysis of historical data.

Control charts

The SPC control charts offer a visual comparison of the current variation in a critical process parameter (referred to as *X* in the subsequent discussion) with its short-term and long-term historical variation. *X* may be the result of a periodic lab test, a measurement signal from a process control sensor, the error in a process control loop, or almost any other quantifiable characteristic.

Two major charts are usually used in conjunction: the *X*-bar chart and the *R* chart. Together, they are referred to as a Shewhart Chart.

The *X*-bar chart shows how the average value (*X*-bar) for a subgroup of individual *X* values varies with time.

As an example, let us say you make widgets that must weigh within 1% of a specified amount. You turn out 10 widgets every half hour, and weigh each one. For purposes of SPC, you group the widgets into subgroups of 10, and plot the average weight (*X*-bar) of the subgroup against time, one *X*-bar value for every

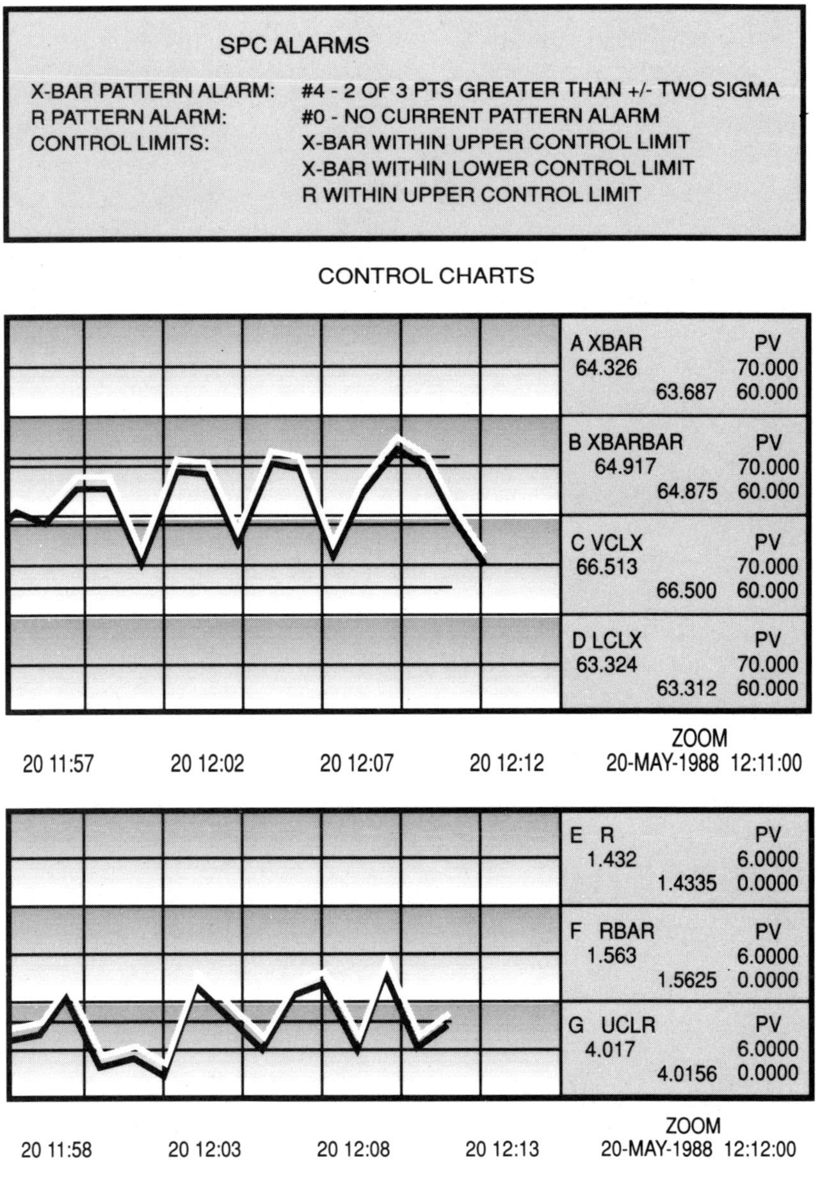

Figure 8-1. Typical control chart

half hour. The variation of *X*-bar with time is a quantified indication of the long-term variation of the widget weight, the process parameter *X*.

The *R* chart shows how the range of the individual *X* values within each subgroup varies over time.

In our widget example, the *R* value for each subgroup would be the maximum widget weight in the subgroup minus the minimum widget weight in the subgroup. The change in *R* over time is a quantified indication of the short-term variation of *X*; in this case, how much the widget weight can vary within a half hour.

Looked at together, the *X*-bar and *R* charts tell much about a process. For example if the shift average for an hourly lab text (*X*-bar) stays relatively constant, but the range (*R*) of test values throughout the shift is much larger during the third shift, a sampling/testing procedure problem might exist on the third shift. On the other hand, a relatively constant range for the shift test values (*R*), with a significant change in the shift average (*X*-bar) for the third shift might suggest a substantial environmental influence on the process during the third shift, which could be compensated for. In yet another scenario, if *X*-bar-bar (the average value for all the *X*-bar values; e.g. the weekly average of lab results) changes over a period of several weeks or months, process equipment might be wearing or fouling, or a laboratory reagent may be degrading.

Abnormal variations and SPC alarms

When you look at the Shewhart Chart you see upper and lower control limits (UCL and LCL) on the *X*-bar chart and upper control limits only on the *R* chart (since the minimum amount of change between samples can be considered zero). The control limits represent three standard deviations above or below the average. According to the laws of statistics, 99.7% of all results should fall within these control limits *if* all variation is random and there are no special causes.

When a process is operating normally, the values of X-bar and R will vary randomly between the UCL and LCL. Figure 8-1 shows how normal, randomly varying results should be distributed in the region within the control limits, which is divided into zones of equal width.

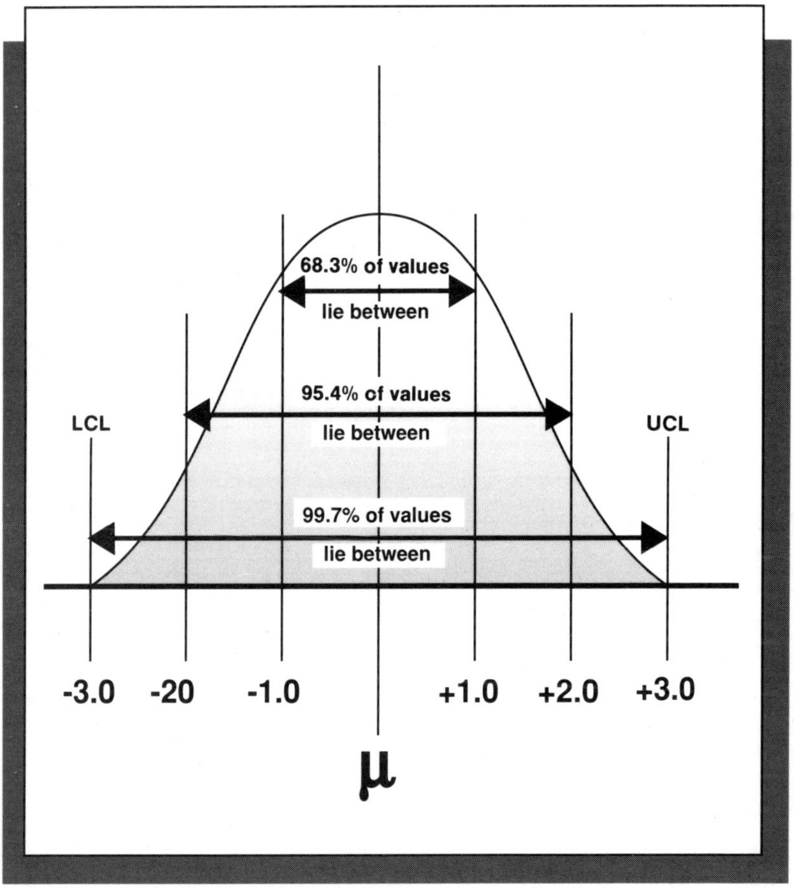

Figure 8-2. Normal probability distribution

Going back to our hourly lab test example, if *X*-bar started falling outside the UCL on the third shift at about the same time that a new technician rotated into the lab for that shift, the cause of the situation would be clear.

Erratic trends on an SPC control chart indicate normal process operation, whereas the same appearance on a typical process control chart would be cause for alarm. In SPC, it's alarming when values don't vary erratically—or when they fall outside control limits. If *X*-bar or *R* start to vary in a nonrandom way (falling outside control limits, flattening out, cycling or ramping up or down, consistently staying outside of the C zones, or exhibiting some other definite pattern), something within the process has changed significantly and the situation should be investigated immediately.

There are two types of tests for abnormal variations:

- Out-of-limits alarms indicates that values are falling above the UCL or below the LCL.
- Pattern alarms indicate unusual, nonrandom pattern occurring, often within control limits.

On-line SPC systems continually prepare charts and compare the data against certain patterns to test for alarm conditions, because identifiable patterns for *X*-bar or *R* within the ABC zones indicate that some special cause is repeatably influencing process operations. Figure 8-2 shows typical SPC alarms.

As long as *X*-bar values vary randomly around the grand average (*X*-bar-bar) and stay within the control limits, the process is "in control." If the control limits are within the manufacturing specification limits for the process, the process is "capable." Obviously, this is highly desirable—it takes capable processes to meet customer specifications. If you improve the process to narrow the distance between the upper and lower control limits, you "improve the capability" of the process.

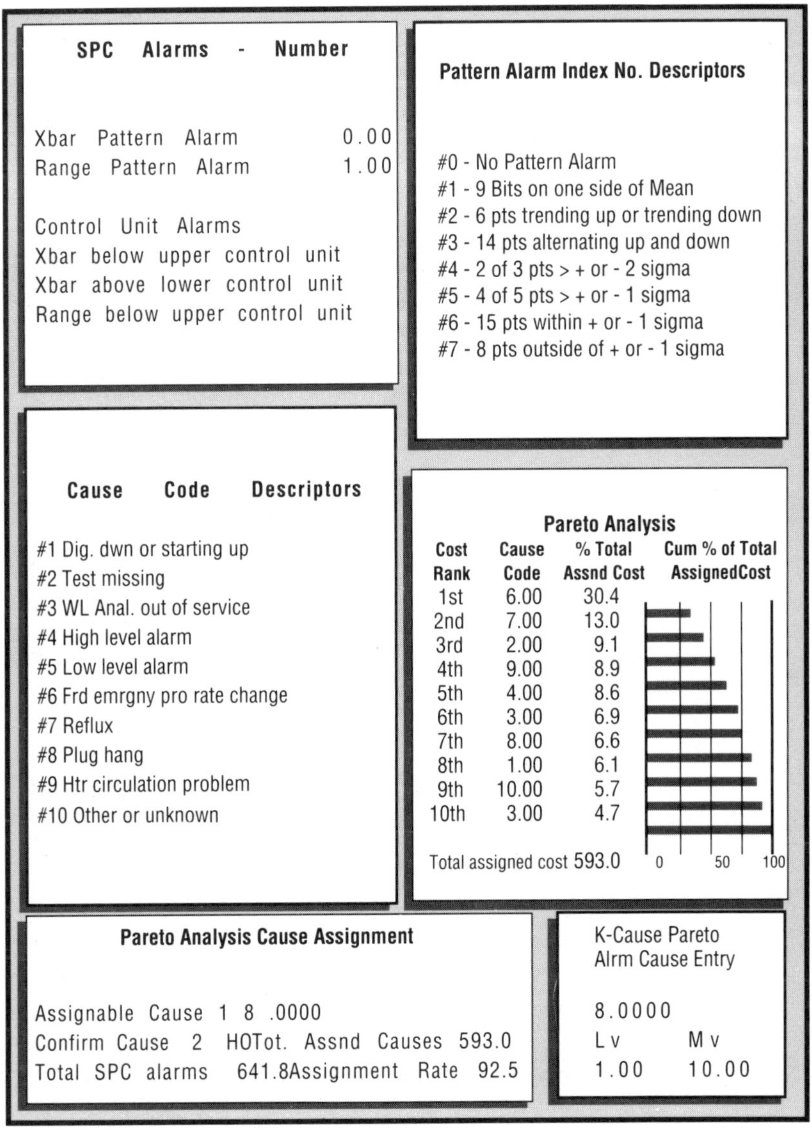

Figure 8-3. Typical SPC alarms

Pareto analysis

Vilfredo Pareto, a nineteenth century Italian economist, is credited with the well-known 80/20 rule. In an SPC context, this predicts that 80% of the abnormal variation in a specific process parameter will be produced by only 20% of the possible causes. This becomes an important factor in improving process capability, because it means that elimination or control of the correct 20% of the potential causes can solve 80% of the abnormal variation problems.

To carry out a Pareto analysis, each abnormal variation must be assigned to a specific special cause and weighted by a cost factor that reflects the cost of not eliminating the special cause. As data accumulates, it is easy to attend to those causes that are the most common and offer the highest economic reward.

The results of a Pareto analysis may change over time. The potential causes in the top 20% category may change with the seasons or with elapsed time since the last maintenance shut-down. And as process operations are improved, many potential causes of abnormal variation will become less frequent or be eliminated and different causes will move into the top 20% category.

Closed-loop SPC

When SPC is used in an on-line environment, it provides an "open-loop" operations tool to identify when some action is required due to variation in a process parameter. Long-term use of SPC may reveal that specific and predictable actions are repeatedly required to eliminate recurring specific special causes. When this repeatability becomes apparent, you can automate corrective actions, thus achieving closed-loop SPC. Corrective actions might include starting or stopping a motor, opening or closing a routing valve, starting an automatic sequence, or biasing the setpoint for a regulatory controller.

Cause and effect correlation study

This is strictly an off-line SPC tool that analyzes how process variables interact with each other, either one-at-a-time or synergistically. A cause/effect correlation study attempts to identify a significant historical cause and effect relationship between an X-Y variable pair, where Y is the critical process parameter that may vary as a function of an independent variable X. For example, the effectiveness of a catalyst (Y) may vary directly according to the temperature at which the reaction is carried out (x).

By constructing an X-Y graph, an SPC analysis creates a "scatter diagram." If there is a noticeable pattern of data points in the diagram, the relationship between X and Y can be quantified using any of several regression techniques.

The results of cause and effect correlation studies are valuable in fine-tuning a process design.

Process capability study

In SPC, the capability of a process refers to the natural variation of a process parameter when there are no special causes that can be attributed to variations. The quantitative indication of the capability of a process are the upper and lower control limits of the X-bar chart for a process variable.

There are three steps in quantifying natural variation and determining the capability of the process:

- Review historical control charts for the particular process parameter to identify variables that fall outside of control limits.
- Eliminate data for subgroups that have a specific cause documented for the abnormal variation.
- Calculate process capability limits (alias control limits) using the edited subgroup data.

SPC in an on-line environment

With a basic understanding of how SPC functions, it is easy to see the advantages of using it in an on-line environment, where undesirable process variation can be identified and rectified before major shifts in product quality and/or process economics are experienced.

Early users of SPC simply issued graph paper and pencils to the operators running a process. Every few months the data was run through a computer program to analyze process capability and look for ways to improve it. While useful in upgrading process design, this off-line approach misses a lot. Off-line SPC control charts often identify abnormal variation long after it has passed, making diagnosis difficult if not impossible. Only on-line SPC can prevent process operating problems from becoming product quality and/or economic problems.

The most advanced and effective way of using SPC is to plug it right into the process control system so that it becomes another operations tool, ensuring products that meet specifications on an hour-by-hour, day-by-day basis. Today, DCS controller modules can cost-effectively provide control charts, automatic SPC alarms and Pareto analysis. These controller modules can transmit the reduced data and alarm messages across the DCS communications network to one or more CRT operator interface units.

In addition to seeing standard process trend charts, operators can also call up real-time SPC charts on their CRT displays. And they can respond to SPC alarms in much the same way they might respond to standard limit alarms. This is a new approach to SPC, and few process control suppliers offer it as part of their packages. Nevertheless, it can be extremely useful and cost-effective. The conventional method of process control is to maintain alarm limits and notify the operator when values exceed the limits. On-line SPC takes this one step further—alarms can be issued *before* limits are exceeded, based on identification and analysis of patterns in the control chart data.

For example, if you are monitoring a critical high-pressure line and the pressure reading stays constant and within the control limits, no standard process control system will ever issue an alarm. But if SPC is tracking that pressure signal and is accustomed to seeing a normal amount of random variation due to pump and valve action, for example, a pattern analysis program would instantly detect a constant flat signal as abnormal and call attention to it. Only on-line SPC can recognize this type of potentially dangerous situation.

On-line SPC also permits an operator to search for a special cause while the condition still exists, so the operator can eliminate or perpetuate the condition to increase operational profitability. This approach can avoid a long production run of off-specification material.

With a DCS implementation, SPC calculations and alarm processing should be distributed throughout the control system. These functions should be located in the controller modules to minimize the burden on the communications network. Distribution of SPC tools brings the same advantages to SPC that distribution of the control functions brings to a process control system.

Adaptive control limits

The normal variation inherent in a process may change as a result of long-term operational improvement or degradation. If the control limits for an on-line SPC control chart represent the normal variation observed in the too-distant past, they may be either too wide or too narrow for the process as it currently operates. This can result in the operator missing situations where the current variation is well beyond the actual normal variation of the process. It can also result in operator response to situations where there is no identifiable cause.

To avoid both false alarms and operator complacency, the computing system should accommodate gradual process changes by automatically adjusting the control limits for on-line SPC control charts. As process operations improve from an SPC perspective, a reduction in normal variation will narrow the control limits and the grand averages will move towards the most economic limit of the manufacturing specification.

What to look for in an on-line SPC system

The following is a checklist of features that are available today, and which you should consider when you choose system for implementing SPC in your environment:

- Complete integration of on-line SPC displays with process control displays.
- Continuous scanning of SPC variables for control limit violation and identifiable patterns.
- Immediate alarming of abnormal process variation to the operator.
- Automatic logging of SPC alarms and cause assignment.
- Automatic compensation for product/grade changes.
- Historical data security. Critical SPC historical variables should be saved in controller nonvolatile memory to prevent loss from electrical power outage.
- Minimum communication load on the process control system. SPC data processing should be located in controller modules where the required data already resides for process control purposes.
- No impact on execution of process control functions. SPC data processing should be segmented from process control functions and executed separately.
- Easy expansion to accommodate multiple process parameters within the unused memory of a controller module.
- Easy modification to provide various types of control charts and pattern tests.

- Easy expansion to accommodate additional pattern tests, products or grades, plus potential causes of abnormal process variation.

These features will help ensure low-cost implementation of on-line SPC functions, fault-tolerant access to critical operations information, and incremental addition of on-line SPC tools to existing control systems.

Implementing on-line SPC with a DCS

The low-cost, incremental expansion capability of a DCS permits low-cost incremental implementation of on-line SPC tools. An implementation sequence might follow a pattern like:

1) First, you modify several DCS controllers to provide on-line control charts and SPC alarms for the most critical process parameter of each major process unit.

2) After a sufficient process history accumulates for on-line calculation of reliable control limits, you institute on-line Pareto analysis, finding each special cause and entering it into the Pareto database. You can estimate the relative cost impact for each cause as you identify it, and adjust it as necessary when new causes are identified.

3) After you have identified enough causes to be able to see trends, you can aggressively eliminate or minimize the most costly 20%.

4) Repeat steps 2 and 3 until the normal process variation is roughly two-thirds of the manufacturing specification tolerance, or until no specific causes can be identified or quickly eliminated.

5) Adjust process control setpoints to shift the grand average closer to the most economical limit of the manufacturing specification without an on-line control limit crossing the specification limit.

6) Monitor process economics to quantify the profitability impact of these process improvements.

7) Carry out off-line cause-and-effect correlations to establish reliable relationships between easily obtained process measurements and difficult-to-measure process parameters.

8) Modify process control strategies to use easily obtained measurements for indirect closed-loop control of the difficult-to-measure process parameters.

9) Repeat steps 2 through 8, justifying additional effort by anticipated profitability improvements.

10) Conduct an off-line process capability study with data collected over a long period of time (e.g. a year) to quantify natural process variation.

11) Manage on-line control limits toward the limits of natural process variation until the economic optimum is achieved and further increases in profitability are balanced by increased costs of additional improvements.

If you apply the Pareto Principle to an SPC program itself, you can see that it should be possible to achieve roughly 80% of the benefit with only 20% of the effort.

In actual implementation, we have seen significant economic benefits from on-line SPC, ranging from $10,000 to $1.4 million a year.

Expert systems

The function of an expert system, the most common term for the current generation of artificial intelligence, is to serve as a consultant, designer, monitor, problem solver, and/or tutor. Expert systems approach these tasks by taking a structured approach to data, information, knowledge, and wisdom.

A structured approach

We covered this briefly in Chapter 4, but it won't hurt to review it again:

- *Data* is the raw number—a pneumatic or electrical signal from the sensor. By itself, it is virtually meaningless.
- *Information* is data that has been checked for validity, scaled, linearized, converted to engineering units, and perhaps checked against upper and lower limits. It means more than data, but still doesn't tell us what to do.
- *Knowledge* is information that has been combined with operating and design experience, control algorithms, and technical objectives to yield some understanding about its relevance. Most expert systems work from a knowledge database.
- *Wisdom* is knowledge that has been combined with external constraints, a preset strategy, skilled assessment of inferred platform states, a forecast of future conditions, plus the overall operating strategy of the platform. Wisdom is the output of an expert system and can be the guiding force for both individual processes and the integration of multiple processes into a complete facilities automation system.

The inference engine

In most expert systems the mechanism that applies the rules to imitate reasoning is called an "inference engine." It is a programmed procedure for combining relevant material from the knowledge database with information from a specific process to reach a conclusion to solve a specific problem.

An expert system is constructed differently from conventional software. Conventional systems use "hard-coded" logic that has to be changed any time new decision-making information comes in. In expert-system software, new information gets added to the system in the form of "rules," without changing the basic structure of the software. The key attributes of an expert system are:

- Separation of rules from reasoning. The inference engine is separate and distinct from the rule base, so rules can be changed, and inference engines can work on any set of rules presented to them.
- Transparency of reasoning. The expert system should be able to reveal its reasoning to its human users. It should not be an impenetrable block of code that presents solutions "deus ex machina."
- Fuzzy reasoning. While this may not sound like an advantage, it is. Engineering reasoning rarely proceeds in yes or no terms—it's usually a matter of degree. The expert system also has to deal with fuzzy data, terms like "high," "low," and "moderate."

Expert system in the on-line environment

Today, expert systems are being used in applications like drilling, well control, pipeline scheduling, refining, dynamic positioning, and many others. Computer vendors are pouring resources in as though artificial intelligence is to be the next major computing breakthrough. IBM now has more people working on expert systems than on database management, and Digital Equipment Company uses the technology in nearly every aspect of their business.

Until recently, most applications of expert systems have run in an off-line mode because of performance constraints. Expert systems bite out big chunks of computer time. The difficulty of providing a common interface between process control systems and expert systems is another reason that few expert systems have appeared in a real-time environment. Expert systems have traditionally run on expensive, large mainframe systems, while process control systems use smaller and simpler microprocessors.

Yet the advantages of on-line expert systems are obvious: smart controls, alarm interpretation and startup/shutdown advisories, to name a few. And today we are seeing products that link expert systems to fully integrated process control systems.

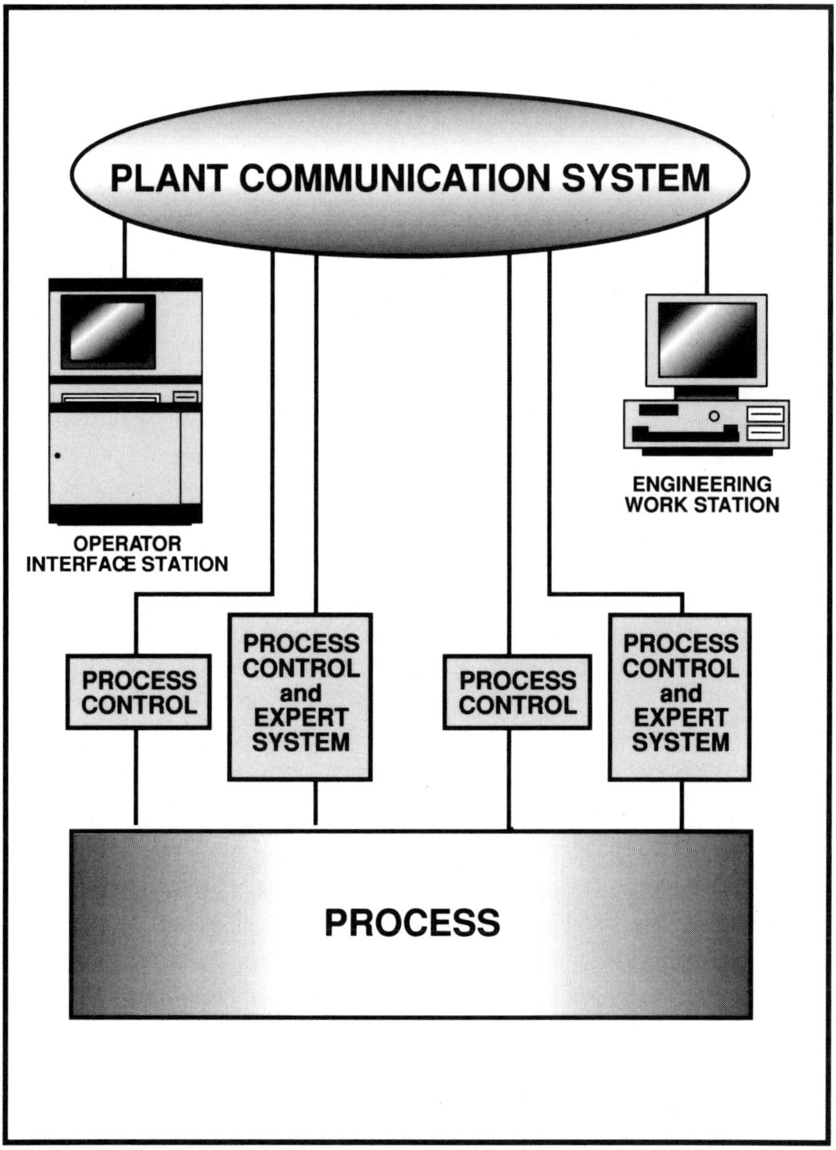

Figure 8-4. Embedded expert system architecture

One vendor offers an expert system that is actually embedded within a process control system. It includes a knowledge workstation that interacts with the human expert in much the same way that the engineering workstation we described earlier works with a process engineer. It uses an interactive approach with simple English statements in the form of "if. . .then" rules, so the person entering the knowledge does not have to be a computer expert too. Once knowledge is entered, the workstation translates it into software modules that run within the distributed control system.

Within the expert process control system, all time-sensitive functions, such as data time tagging, averaging, real-time interfacing, and closed loop control continue to execute within the control system. And all communications between the process operator and the expert system are handled by the standard operator interface of the process control system. This expert system can be distributed among different locations within the process control system to improve its robustness and real-time performance. And it can be interfaced directly with currently installed distributed control systems without special hardware or software requirements.

Advantages of an on-line expert systems:

Having an embedded expert system is like putting an expert human on call 24 hours a day, 365 days a year. The expert system can:

- Monitor system performance and operator actions and respond to problem situations.
- Reconfigure itself based on auto/manual operating modes.
- Modify setpoints and tune parameters based on current process conditions.
- Walk an operator through a process startup or shutdown, reacting to events that happen in real-time.
- Diagnose process problems based on events and trends that are taking place, which makes expert systems indispensable in predictive maintenance applications.

- Predict problems based on SPC and other data and help in predictive maintenance scheduling.

One of the most important uses for expert systems is in the area of safety: An expert system can analyze a complex series of interrelated alarms, figure out what is happening, and take corrective action in the time it would take a human just to recover from the initial panic.

Expert systems provide a unique opportunity for operators, maintenance people, and supervisory personnel to take advantage of the most complete human knowledge available without paying for actual humans.

Summary

In this chapter, we have discussed Statistical Process Control and Expert Systems in some detail as the two principal tools for optimizing process control and effectively managing an automated facility. In the next chapter, we will look at particular applications that may be automated.

Chapter 9

AUTOMATION APPLICATIONS

No two automation applications are ever exactly alike. While we will not go into all the details here, we will attempt to give a broad outline of the possibilities. The general categories into which offshore platform applications fall as are follows:

- Production control—oil
  - Wellhead
  - Enhanced oil recovery
  - Separation
  - Pipeline
  - Reservoir management
- Production control—gas
  - Wellhead
  - Dehydration
  - Separation
  - Compressor control
  - Gas turbine control
- Maintenance systems
- Safety systems
  - Gas detection/fire control
  - Emergency shutdown systems
  - Structural analysis
- Support systems
  - Power generation
  - Energy management
  - HVAC
  - Fresh water

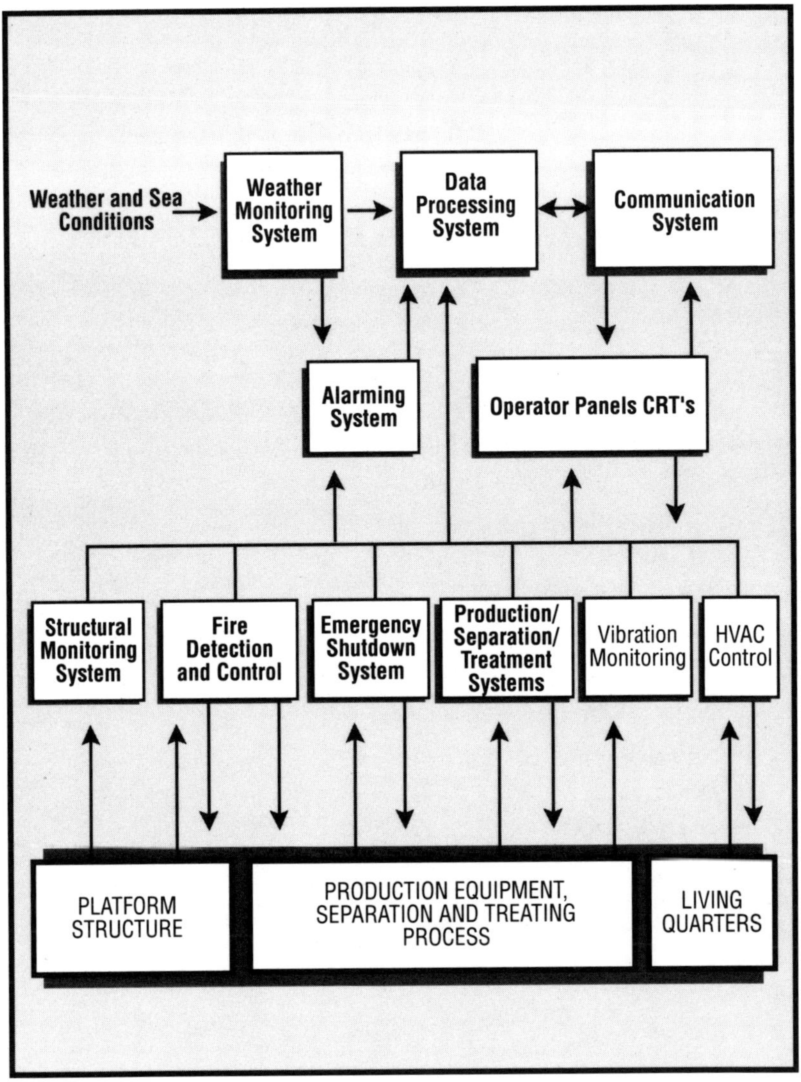

Figure 9-1. Production platform subsystems

Figure 9-1 shows the relationships between the platform, equipment, people, and the various monitoring and control subsystems in an integrated approach to production platform automation.

Automation may be applied to a single platform or to multiple platforms connected to shore. In a single-platform environment, the major emphasis of automation will be to replace existing controls with distributed controls and integrate the control functions.

Platform-to-shore communications can extend the capabilities of the system by allowing shore-based operators to operate the platform (for example, during storm conditions when the platform is unattended), and by providing more extensive data analysis.

In the most sophisticated systems, a central manned platform will control operations on multiple unmanned platforms that are all part of the distributed control system. Most of the systems like this also include a communications link to shore.

Automation can be applied in other areas that relate to offshore production:

- Vessel automation
- Drilling
- Training and simulation

Production control

The type of applications considered under production control vary depending upon the location of the well and the type of crude oil. On a fixed platform in the Gulf of Mexico, a vast pipeline network eliminates problems of separation and storage—crude oil goes directly from wellhead to pipeline. Here the major issues are being able to shut in the well during hurricanes, and protection against blowouts and compressor surges.

In floating production systems far from pipeline distribution networks (for example, in the Far East) other issues arise. Here operators deal with subsea completions, wellhead separation and treating; storage and transfer; and handling of by-products, whether that involves water treatment and reinjection or utilities that use the gas generated by the well for power and heat.

Because both gas and crude oil share many elements of production control systems, we will cover both in this section.

Subsea completions

Subsea completions are especially applicable to small reservoirs, reservoirs in deep, hostile waters, unusually shaped oil or gas fields, and drilling locations that cannot be reached by directional drilling. This technology allows oil and gas fields to be brought on-stream in a very short period of time, often at considerable cost savings.

A typical subsea production plan consists of a surface production facility to separate oil, gas and water; a production riser to bring the oil and/or gas to the surface for processing; the subsea well itself, complete with wellhead, production tubing, Christmas tree, flowlines between various stages within the subsea production system; and a control system.

Subsea wells may be completed individually as satellite wells, or they may be completed in groups on a template. The template can be simple, without commingling, or it may house a manifold where the products from several wells commingle before going to the production riser. Clustered subsea completion systems integrate an entire field development program by combining wellheads, casing hangers, and all associated subsea drilling and production equipment on the leveling template. One advantage of this approach is that an operator can drill and complete production wells while the permanent production facility is under construction.

Control systems for subsea completions fall into two major categories: hydraulic and electrohydraulic.

Hydraulic systems are the oldest, and were for along time the only truly reliable solution. There are two main types of hydraulic systems:

- Direct hydraulic systems use one hydraulic umbilical for each subsea function to be controlled. Hydraulic pressure from the platform directly operates valves, chokes, etc. Although very reliable, the use of one hydraulic line for each function makes this approach quite cumbersome in all but the simplest installations.
- Discrete and sequenced hydraulic systems improve the response time of hydraulic systems and simplify their use by storing hydraulic pressure in subsea accumulators within the subsea control package. A signal from the surface directs the accumulator pressure to the desired function, and a single hydraulic supply line maintains accumulator pressure.
- A discrete system uses a hydraulic signal from the surface to shift a pilot valve to direct accumulator pressure. This system also requires one control line per function.
- A sequenced system minimizes the number of control lines. In a pressure-actuated system, pilot valves operate at preset pressures, so that half a dozen pilot valves can be operated by one hydraulic control line. In a time-sequenced system, one control line can be used to step through all control functions in a preset sequence.

While hydraulic systems are more flexible, they still suffer severe limitations. First, when the distance is greater than 2-3 miles, the long period oftime required for a hydraulic signal to reach the subsea location and actuate a control may make the system unsafe. Second, hydraulic systems do not lend themselves to monitoring temperatures, pressures, and equipment status, which is almost mandatory for subsea templates. And third, pure hydraulics offer no form of local intelligence at the subsea location.

The solution most used today is some form of multiplexed electrohydraulic system. In most cases,the hydraulic umbilical still maintains pressure in the accumulators (although it is possible to eliminate this umbilical by installing a small, electrically-driven pump subsea to maintain pressure); but signals are transmitted to and from the subsea control unit by other means—electrical wire, acoustic transmission, or fiber-optic cable.

Today's subsea completions are quite complex. In the North Sea, the Balmoral Subsea Control System handles a central fourteen-slot template with ten production wells and three manifolds, three satellite production wells, six water injection satellite wells, and a floating production vessel moored over the template.

The design of the Balmoral subsea system provides for control of up to 300 subsea tree and manifold valves, subsurface safety valves and chokes. The system can also monitor up to 130 subsea data points including downhole pressure, choke positions and differential pressures, wellhead pressures, and certain manifold header pressures.

A total of eighteen control modules are in use on the Balmoral template: ten production tree modules, three satellite water injection tree modules, three satellite production tree modules, and two template riser modules. The use of one module per production tree helps assure maximum production at all times.

Because the wells have widely divergent production characteristics and produce into commingled headers, each well has sensors to monitor upstream pressure, the differential pressure across the choke, and the exact choke position. Operation of each choke is controlled by the control system on the floating production vessel. The control system also provides real-time operator schematics of virtually all aspects of subsea operation.

Current technologies allow for distances up to about 12 miles, but technologies to expand that range up to 100 miles or more are under development. The key problems to be addressed include data and command transmission, local subsea intelligence requirements for slug catcher and pump regulation, and control and cost of cabling for the subsea pump power supply.

New developments are also needed in subsea fluid handling methods—a breakthrough in multiphase fluid transmission could cut line costs and gas transmission problems.

As technology continues to expand, we should see greater use of distributed control systems in the subsea environment. Distributed processing units on the sea floor offer a level of control that cannot be achieved from a distant surface location.

Wellhead

The operation and status of each individual well within a field can be continuously monitored, and performance and malfunctions reported to remote operations personnel. Automation can be applied to virtually any type of well: free-flowing, pumped, gas lift, or water injection, for example.

- Remote control of production or injection chokes can maintain flow rates. Well flows can be allocated across an entire field and optimized based on selected field performance criteria.
- Wells can be brought on line or shut down remotely, in a coordinated manner.
- Monitoring the wellhead pressure and temperature of free-flowing wells can indicate well performance. Choke settings can be changed remotely to maintain the desired production.
- For gas lift wells, the gas lift flow rate to each well can be monitored and controlled to optimize production. Automated antisurge control can minimize excess compressor recycle gas, protect the compressors against pressure surges.
- "Off" conditions and pump valve failures in producing wells can be identified and alarmed to the remote operator.
- Water injection into the injection well can be monitored and controlled remotely.
- Records of equipment maintenance schedules and malfunctions can be maintained by the system to minimize lost production and aid in problem diagnosis. For example, use

of remote sensors to generate the dynamometer card data for mechanical sucker-rod wells can provide an efficient tool for failure analysis before maintenance personnel are dispatched to a well failure.

- An automated system should provide three levels of safety protection: alarming for operator attention; automatic corrective action (shutdown);and pressure relief mechanisms to protect against blowouts caused by excessive well pressures.

Separation

In wells producing low quality crude oil, the separation process is critical to producing marketable crude. In the Far East, in particular, the crude has a high sulfur content and is a poor mix of gas and oil; the cost to refine and transfer it is higher than anywhere else in the world. Inconsistency can reduce the marketability of Far East crude still further.

The separator is the series of units that separate water, gas and oil. Many fields have a multiple production separator system (shown in Figure 9-2) to separate wellhead streams coming in at different pressures and stabilize the oil by removing light ends.

Depending on the flowing pressure of a well, the production stream can be directed to the high-pressure (HP) separator, the intermediate-pressure (IP) separator, or the low-pressure (LP) separator. Each separator operates as follows: As the stream passes through, gas rises and passes out through a mist eliminator that removes oil and water droplets. The gas is metered and directed to a pipeline or to the gas lift injection system. Liquid settles out within the separator, into an oil and a water layer. Within the unit, the level of the interface between the two layers is controlled by the rate at which oil and water are removed. As they are removed, both oil and water flow are measured by turbine meters or positive displacement meters, after which they travel to their respective treating systems.

In a multiple separator system, the operating pressure of each unit is controlled by the gas back-pressure control valve on each. Oil coming in ultimately flows through all separators operating

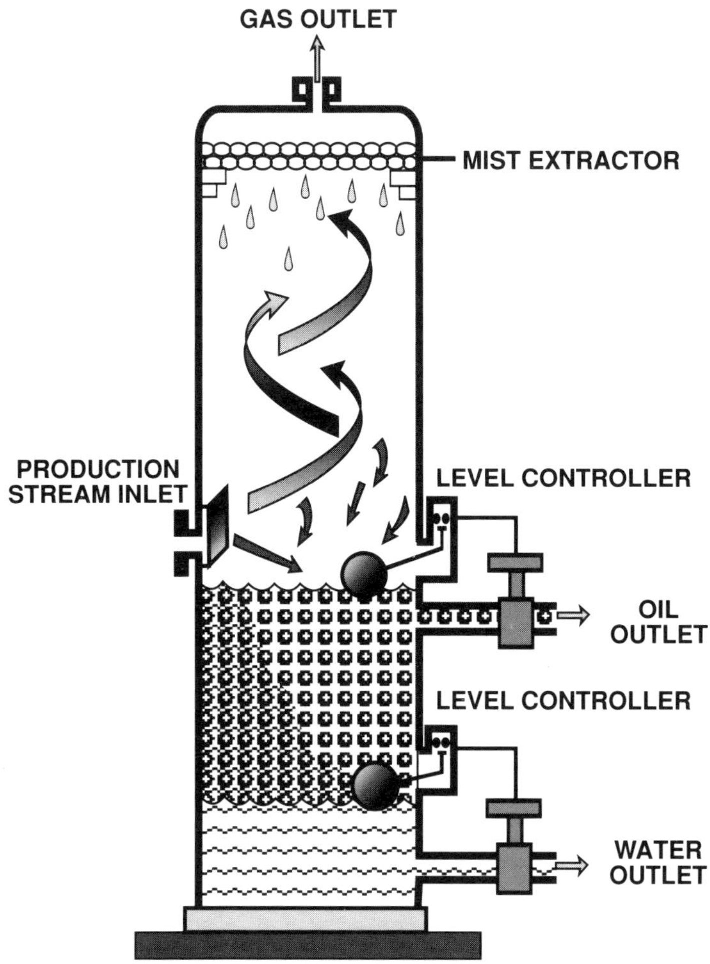

Figure 9-2. Typical oil separation unit

at or below its initial pressure. Thus oil coming in first to the HP separator flows from that unit to the IP and then to the LP separator. This process allows most of the light ends to vaporize, which stabilizes the oil for future treating and transport.

Virtually all production separators also require a test separator, where streams are diverted periodically in order to measure the volume of various constituents, important for determining operating parameters of the main separators. Some areas where automation can be applied include:

- Remote monitoring of unit temperature, flow, and pressure to maintain optimum operating conditions within each separator.
- Monitoring wellhead pressures and matching separator pressures as appropriate. Measuring gas flow and pressure through the gas drying system and to the compressors to anticipate and prevent surge conditions.
- Monitoring the oil/water interface level to prevent upsets and provide constant flow of oil and water out of the units.
- Automatically switching wells to the test separator for analysis, in order to determine the best operating parameters for the main separator.
- Providing a safety system that would: alarm operators of out-of-limit conditions; provide an automated, orderly shutdown if required; and protect against excess pressures with pressure relief valves.

Among the most attractive prospects in separation automation are multiphase flowmeters now under development. This type of meter offers promise of a near-exact reading of the percentage of sand, oil, water, and gas being produced. Such a remote sensing device in the wellhead could supply the information to optimize the separation and treating process, and, even more important, could eliminate the need for a test separator and all the piping that goes with it.

While a working multiphase meter may still be some time off, mass flowmeters are now being used to arrive at an approximation of stream content. For example, in the Far East, density and sulfur content have a direct relationship to each other and to the quality of the crude. Knowing these factors can enhance the separation process.

Oil treating

Oil from the separator goes to untreated wet oil storage, because it still contains quite a bit of emulsified water. To remove it, the oil is treated with a demulsifier, heated, and sent to the oil treating unit.

The first section of the oil treater operates much like a three-phase separator (gas/oil/water), handling the additional gas and water released by heating with the demulsifier. In the second section, water droplets coalesce on an electrode grid (electrostatic demulsification), and fall to the bottom of the tank.

Produced water goes to the water treating unit, and dehydrated oil is analyzed for basic sediment and water (BS&W) content. If BS&W is less than one percent, the oil goes to the dry section of the oil storage tank. If it is above one percent, it is sent to a wet oil tank to await a second trip through the treating unit.

Automation can be applied to the oil treating unit in several areas:

- By monitoring the temperature of the oil to be treated, the amount of demulsifier added, and the resulting BS&W after treatment, operating conditions can be optimized to help ensure that all treated oil falls below 1% BS&W.
- Monitoring the level in the wet oil tank can signal when production must be shut down to allow this oil to be reprocessed. An automated system can handle this shutdown and switching in an orderly and efficient way.
- By applying on-line Statistical Process Control to the treating process, trends can be monitored to prevent out of limit (greater than 1% BS&W) conditions. This can reduce the amount of wet oil produced and eliminate production shutdowns needed to handle reruns.

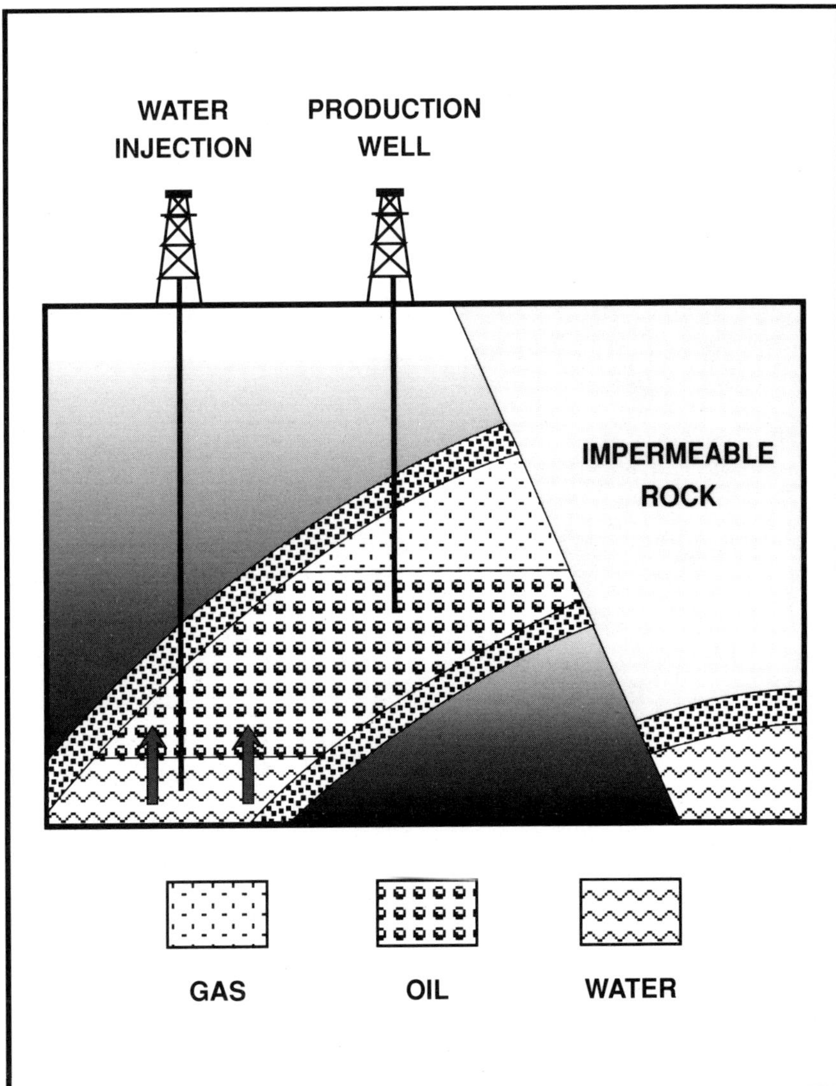

Figure 9-3. Typical water injection system

- Safety monitoring can alarm operators when control limits are exceeded, conduct emergency shutdowns, and provide relief for overpressure situations.

Water treating

Just as separated oil contains water, so separated water contains some oil. The first step in removing this oil is normally a corrugated plate interceptor or other type of skimmer, which reduces oil content in produced water from about 3,000-10,000 ppm to 150-600 ppm.

If water must be cleaner than this, it goes through a gas flotation cell, where gas is bubbled through the water. Dissolved oil tends to stick to the gas bubbles and rise to the surface where it can be skimmed off. This process typically reduces oil content of water to less than 50 ppm.

Automated systems may be used to monitor tank levels, equipment alarms, shutdown and performance. In addition, keeping maintenance records in a computerized database can help identify repeating problems and failures and thereby reduce maintenance costs.

Water injection

Water to be injected back into a formation must be clean: free of oil, oxygen, and biological contamination. A typical system appears in Figure 9-3.

Injection water is first strained to remove particles greater than 100 microns, and a flocculent is injected to aid in the filtration of finer solids.

Filtration generally takes place using downflow vertical filters regulated by a cycle programmer or control system. Two filters are normally on line, while one is active, the other is being backwashed and cleaned by clean water. While filters are generally cycled on a fixed time schedule, two events can interrupt the normal timing: high differential pressure across a filter (indicating clogging), or excess turbidity downstream (indicating breakthrough).

After filtration, the water flows through a control loop to the top of the vacuum deaerator. Here it flows down over a packed bed as a vacuum removes trained air. The level controller in the deaerator controls water flow.

In addition to filtering and deaerating, the water is also treated: Antifoaming agents, biocides and oxygen scavengers are injected prior to deaeration. After deaeration and before injection, the water is analyzed for dissolved oxygen and corrosion. If the water exceeds control limits, it is diverted until operations return to normal.

Water injection pumps then pump the deaerated water into the injection wellheads. A scale inhibitor is generally added at this time, and downstream monitors check for corrosion and biofouling.

Automated controls that provide an indication of trends can help keep this system running smoothly:

- Monitoring the condition of filters and backwashing them as needed, rather than on a preset time basis, can allow filtration to proceed without interruption or upset, possibly eliminating shutdowns.
- Monitoring injection flow rates into the wells can prevent hydraulic shock and possible damage to the producing formation.

Compressor control

Modern centrifugal compressors provide efficient and cost-effective compression of gases to a wide range of pressures. However, due to limitations inherent in the physics of centrifugal compression, the compressors are generally limited to operations between 50 and 100% of their rated capacity. If the flow through the compressor falls below 50%, surge occurs and can cause major equipment damage.

Centrifugal compressors surge when the impeller cannot generate enough head (ft-lb force/lb mass) to keep the compressor discharge pressure higher than the downstream piping pressure. It happens when there is not enough gas mass flowing through the compressor. In this circumstance, the open-vane structure of the impeller provides no check on flow reversals, and gas flows backwards, from the discharge to the suction side of the compressor. When enough gas has accumulated at the suction side, the flow again reverses to reestablish the normal flow direction. And when the temporary excess supply of gas on the suction side is depleted, the flow reverses again. The repeated cycling of gas within the compressor is surge, and it is very detrimental, both to the compressor and to the turbine or electric driver of the compressor.

Surge causes extensive damage to compressor seals, bearings, gears and drive, due to the violent motion of the impeller during the flow reversals and the loading/unloading of the driver. Catastrophic failure of the compressor or the driver (or both) is a significant possibility if surge continues any length of time. Even short-term surge causes excessive mechanical wear of compressor and drive parts.

Automated control systems can prevent surge and significantly reduce maintenance and operating costs. The control system operates a bypass loop that takes gas from the discharge side of the compressor and feeds it to the suction side so that the total gas flow through the compressor is above the point where surge would occur. To avoid surge, most control strategies leave the bypass valve open slightly all the time and set the minimum gas flowrate significantly higher than the actual surge point for the compressor. While this ensures enough head to keep the compressor from surging, it also wastes energy by compressing more gas an is really needed. Advanced control strategies can accurately predict the surge point, so that the bypass valve can be closed in normal operation, and only enough gas to prevent surge is bypassed. The major benefit is a significant reduction in compressor operating costs.

Reservoir management

Using a DCS to monitor producing reservoirs (patent pending) allows oil and gas production to be optimized to the lowest cost over the longest possible production life of the reservoir.

Oil and gas are produced based on three factors: the flow properties of the oil, the size and condition of the reservoir, and the current market price.

Historically, reservoir modeling has been an ultracomplex and expensive computer task reserved for only the largest and richest reservoirs. This meant that small reservoirs were often not produced under optimum conditions.

Computing complexity added another factor to those reservoirs which were modeled: the time between the collection of reservoir modeling data and the final analysis of results took weeks. While this was not a problem during the days when crude oil prices were high, it is a real problem today.

The characteristics of a reservoir do not change quickly, but the price of crude does. And because the price of crude often remains close to the cost of production, a well that is profitable today may be operating at a loss tomorrow. Many wells operate on the borderline of the price of crude, so whether or not production is justified is a question that should be looked at more often and evaluated much more closely.

Fortunately, with the advent of inexpensive computing power and expert systems it's now possible to monitor a reservoir in real time, using a DCS to gather the data in the course of its normal work controlling the wellhead and production. Data from each well can be correlated with similar data from other wells tapping the same reservoir to give a remarkably complete picture of what is happening beneath the surface.

Some of the reservoir parameters that can be monitored and modeled by a DCS include:

- Well pressure. The expansion of the gas cap that sits over the oil in a reservoir is a useful tool in balancing withdrawals from the reservoir. Ideally, the pressure across the reservoir will remain constant if oil is withdrawn uniformly from all areas.

- Gas-to-oil ratio (GOR). GOR maps are developed for a field by posting current GORs for each well and drawing iso-GOR lines. An area with a high GOR will normally coincide with a low pressure area, and indicate depletion of that area of the reservoir.
- Water-cut ratio (WC). As the amount of oil in a reservoir declines, a well tends to produce more water. A map of water cuts from wells in a reservoir can corroborate information on the GOR map.

Reservoir modeling can also be useful in well maintenance. For example, if there is an anomalous high gas producer, it is possible that the gas is coming from another zone that is leaking through to the production perforations. A bottom-hole temperature gradient survey can help locate the problem, after which a cement squeeze workover can seal off the leak.

The overall result of on-line reservoir management would be timely information for management decisions about how best to operate the wells. And once management decisions are made, the DCS can effect the agreed-upon control strategy in real time.

Pipeline

Pipelines may transport liquid, gas, or a mixture of the two. Perhaps the most important automation and monitoring application is pipeline metering for sale and custody, which is both well-understood and widely implemented. A less used DCS pipeline application is monitoring for leaks.

Leak detection

Leak detection for both liquid and gas pipelines can be carried out by monitoring the pressure and mass-flow measurements at the entrance and outlet of the pipeline. Mathematical models can detect leaks occurring suddenly and estimate the location and leakage rate. Tests have shown that gas or liquid leakage rates as low as 1% can be detected.

The mathematical model is formulated from measurement data collected during an operating period with no known leaks. Leak detection monitoring compares measured signals with the model data, using relatively little computing power. The application can be easily implemented using a microprocessor-based DCS.

Predictive maintenance (vibration monitoring)

Vibration monitoring is a relatively new application rapidly gaining acceptance, because those using it have documented savings ranging from thousands to millions of dollars per year.

Vibration is a form of wasted energy caused by mechanical problems in a machine. By measuring the amount of wasted energy, we are actually measuring the severity of the mechanical problems.

The application entails attaching sensors that monitor vibration to machinery with rotating parts. Monitoring vibration can lead to more efficient operation of machinery and more cost-effective maintenance. Even more important, it can extend equipment life and prevent catastrophic failures.

Methodology

Before we talk about vibration sensors, it helps to understand exactly what we are measuring. Figure 9-4 shows a waveform, the movement of a vibrating part through time.

The two things to consider when measuring vibration are amplitude and frequency. Amplitude has three components:

- The peak-to-peak distance is the **displacement** which is usually measured in mils (0.001 inch).
- The **velocity** is the speed with which the vibrating machinery is moving up and down—it is a measure of how long it takes the machine to go from the trough to the peak of the wave.

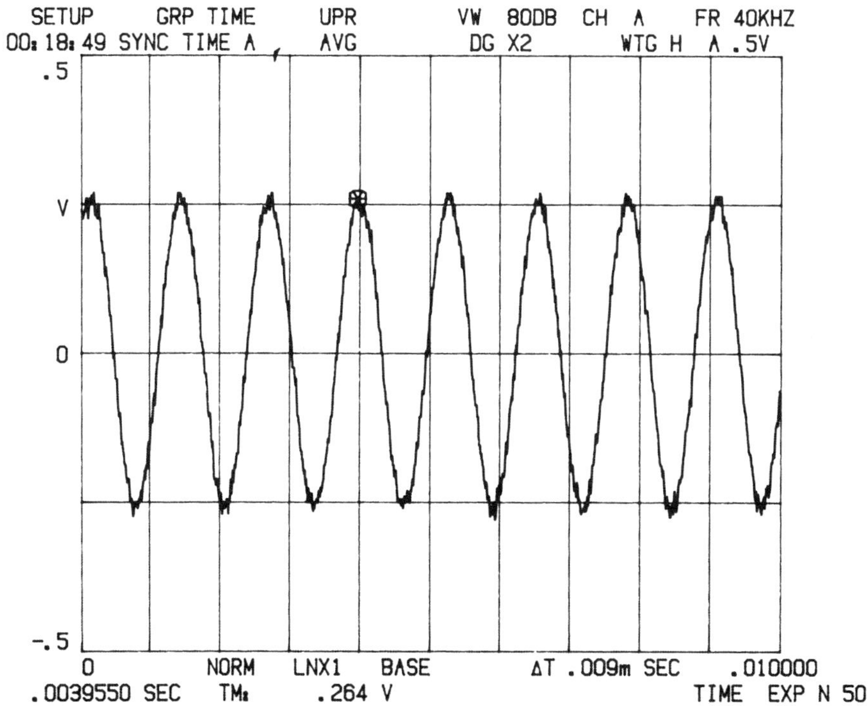

Figure 9-4. Typical waveform

Velocity can be measured as the average velocity, the peak velocity (reached as the vibrating part passes through the normal position midway between the upper and lower limits), or the RMS velocity. Velocity is given in inches/second.

- **Acceleration** provides a third way of measuring amplitude, and it reflects how fast the velocity is changing. Acceleration peaks as vibration screeches to a halt in one direction and starts up again in the other direction. Acceleration is zero at the normal position, when velocity is at a maximum. Acceleration is measured and stated per standard gravity.

Note that all three of these measurements above refer to the waveform Shown in Figure 9-4. They have nothing to do with **frequency** which is another measurement altogether. Frequency is the number of cycles per unit of time, and is usually given in cycles per second (cps) of hertz (Hz).

Displacement, velocity, and acceleration are important because they provide different ways of measuring the amount of vibration. Frequency is important because different problems arise at particular frequencies. Frequency analysis lets you determine hatthe problem is, while amplitude measurements give you an idea of how bad it is.

Vibration monitoring starts with some type of transducer on the machinery, mounted externally or internally. There are three common types of transducers:

- The anticurrent probe measures displacement, which is a good mode of measurement for low-speed operations.
- The velocity sensor measures velocity, an effective measurement for all frequencies of machine operation. Mathematical integration of the velocity reading yields the displacement of the vibration.
- The accelerometer measures acceleration, which is useful for high-speed operation (above 10,000 rpm), for detection of sharp transients, or for detection of faulty antifriction bearings or gears. Integration of the accelerometer reading yields velocity, and double integration yields displacement.

Clearly, choosing the right mode of measurement and the right transducer for a particular job is no trivial task. It depends on the type of bearings, the speed of the machine, the dominant frequencies to be measured, the temperature at the mounting point, cable lengths and strength of signal, relative mass considerations, all too detailed to treat here.

Transducers are connected to a monitoring device that continuously reads vibration information and tests it against preset limits. Alarms are tripped as various preset limits are reached.

While this sounds simple, it may not be so simple in reality. The control system must filter signals to pinpoint the critical frequencies for machine operation while ignoring spurious

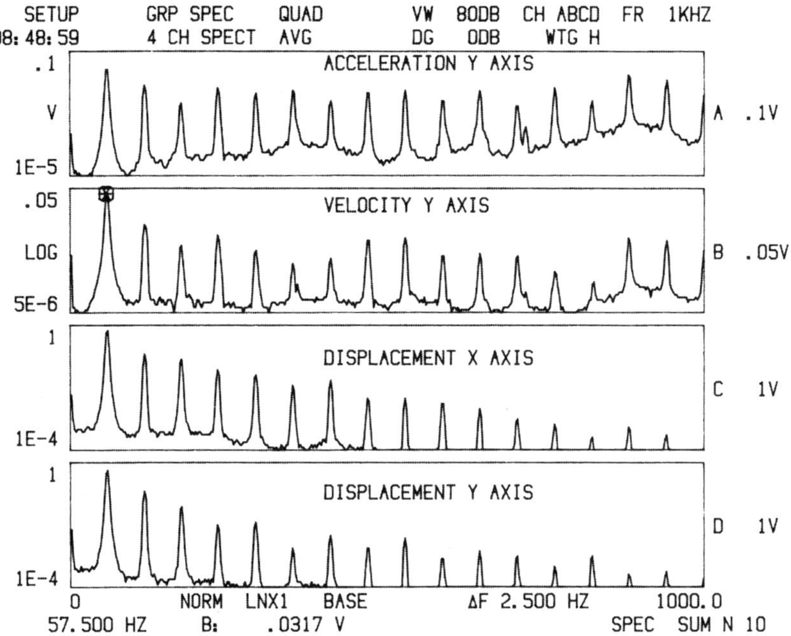

Figure 9-5. Vibration signatures

signals. These may result from wind, waves, and even boats that run into the platform.

Another factor that complicates offshore installations is that space is at a premium. That means that one sensor may actually be monitoring more than one piece of machinery because the equipment is so close together. So offshore vibration monitoring systems must be quite sophisticated. The data collection system must isolate the various frequencies in the vibration spectrum to pinpoint the activity associated with each machine.

This kind of analysis can dig even deeper, pinpointing a problem to a specific part of a machine—a bearing, loose shaft, loose foundation, belt slippage, etc. Some vibration systems can even analyze problems within a bearing, giving details on the performance of the inner race, outer race, and ball shield. Knowing the exact part of a machine that is failing can be important in estimating the amount of time before the problem becomes a disaster.

Vibration monitoring can be carried out on a periodic or continuous basis. Regardless of whether the data is analyzed once a minute or once a month, the important concept is trending. Operational trends provide the information needed to establish predictive maintenance practices and schedules.

It is preferable to monitor continuously, because some failures can happen almost instantaneously, and these failures tend to be the ones that blow things apart. If continuous monitoring is not feasible, you may want to watch trends and take readings more often as values approach control limits.

Of course, even in a "continuous" system, each reading is an average because digital systems scan at a finite rate. Each scan reports an average of the analog values since the last scan.

Sophisticated vibration monitoring systems use the Fast Fourier Transform method to reduce a complex incoming signal (a mixture of waveforms that may include inputs from multiple machine components, multiple machines, the effects of wind and waves, and so on) to simple graphic displays of what is happening at the basic frequencies. You can see not only how overall vibration is changing, but how certain components of vibration are changing.

The graphs in Figures 9-5 and 9-6 show some of the data produced by vibration monitoring systems:

Advantages

The principal use of vibration monitoring is preventive and predictive maintenance:

- A vibration "signature" for new equipment can serve as a quality control check of the equipment as it comes from the manufacturer, and as a reference point for subsequent analysis and maintenance.
- Results of vibration monitoring can be used to schedule maintenance. Work crews can be sent out only when

necessary, and they can be supplied with exactly the parts they'll need for the specific equipment that needs attention. As a side benefit, the morale of the maintenance department will improve.

- By knowing the exact problem within a machine, it is possible to fix it in the minimum amount of time, often without taking the machine down.
- Vibration monitoring helps extend equipment life. If you catch and correct an out-of-balance situation, you extend the

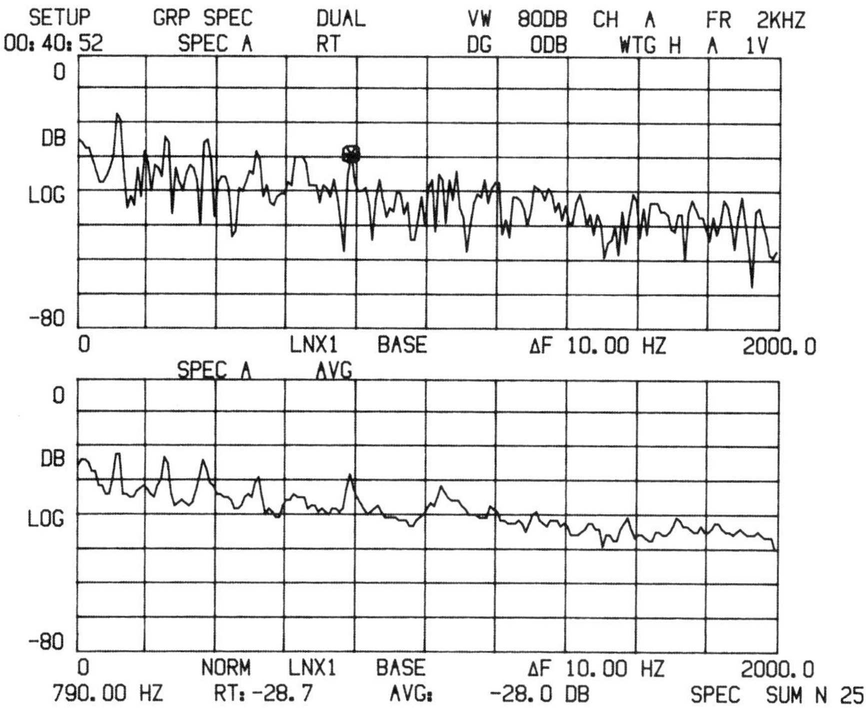

Figure 9-6. Vibration trend analysis

life of the bearings, and prevent a catastrophic failure that could ruin the whole machine or the whole platform.

- Maintenance inventories can be reduced, because repair parts can be brought in based on projected need.

Vibration monitoring systems provide a good area for expert systems. Not only can such a system diagnose problems and potential problems down to the exact part of a machine that is in trouble, it can also provide other benefits. For example, as one of the inputs feeding the control system, vibration monitoring can keep a process running at the optimum rate for the machine. This might be important in ramping a turbine up to speed at a controlled rate to avoid premature wear. Or it might be useful in analyzing the economics of running marginal equipment slower so that it would not have to be repaired as soon.

Cost savings

In general, vibration monitoring applications will pay for themselves in less than one year. The savings fall into several areas:

- Preventing unscheduled downtime. Murphy's Law tells us that if a machine is going to fail, it will do so at the worst possible time. Vibration monitoring prevents unexpected failures.
- Reducing downtime and lost production. Knowing exactly what is wrong makes it easier to fix.
- Maintenance scheduling and reduced inventory. One major oil company with eight refineries estimates that they save millions of dollars a year in maintenance and increased throughput. A major paper mill saved thousands of dollars each month in reduced repair time and lost production.
- Extending the life of the machinery. All users find their equipment lasting longer because they can repair it before it breaks. A mining company in the western U.S. saved $100,000 in machinery replacement costs alone in the first eight months of their program.

- Reducing energy costs. Vibration is wasted energy. Machines that vibrate less require less energy to operate.
- Safety. While this isn't easy to quantify, it can be critically important. Because vibration monitoring indicates when something is about to go bad, life-threatening failures—the sort that can happen when a large gas turbine bearing goes—can be avoided. This can lower both liability exposure and insurance costs.

Safety and emergency shutdown

We have discussed many of the important features of safety systems in Chapter 7, where we talked about robustness, fault tolerance, and redundancy. Here we'll look specifically at fire and gas systems and the emergency shutdown application.

Fire and gas detection

A fire and gas detection system receives data from various HVAC and environmental sensors, including temperatures, smoke detection, gas detection, strain gauges, A/C blower operation, etc. Through a CRT display, the operator can monitor and control the fire pump, damper and fans, Halon and CO_2 systems. The operator should also be able to control watertight doors and hatches to isolate damage areas of the vessel or platform.

Some fire and gas systems are extensive and complex. Such a system may take considerable control action on its own; it may use extensive panelboards and CRT displays in different areas (for example, on the platform and in a shore-based control room); and it may even operate voice synthesizers to tell operators what to do next.

Emergency shutdown systems: damage control

As emergency shutdown (ESD) systems have moved away from simple pneumatic implementations with air pressure holding open a shutdown valve, the concept of emergency shutdown has broadened to one of damage control; anticipating problems and preventing or minimizing the outcome.

Damage control has three phases: quantification of data, taking corrective action, and reconstituting operations.

Quantification of data refers to analysis of the data turned in by sensors. It means analyzing trend data and out-of-limit conditions in the light of everything going on in the process in order to anticipate and prevent major problems.

Taking corrective action means developing a strategy. For example, if you are fighting a fire with seawater, you must take into account the effect that the additional water weight will have on the stability of the structure. Some say that more vessels have been sunk fighting fires than from the fire itself.

Reconstituting operations can be done manually, or it can be designed as a system function. It depends on what level of intelligence you want in the system, and how much self-sufficiency you are willing to give a particular application.

As you can see, the "damage control" approach to ESD requires some form of expert system, particularly in large applications.

Elements of an ESD

Every ESD system, complex or simple, has several common features:

- It must be highly reliable. That includes high-quality components, built-in diagnostics and fault tolerance and appropriate level of redundancy, and, preferably, self-testing capability. The system must operate for years without being called upon to perform.
- It should provide a preprogrammed, rapid, and consistent response to emergencies, 24 hours aday, 365 days a year. No system should rely on the capabilities of a particular operator.
- It should function independently of other instrument and control systems. Within the overall ESD, each subsystem

should stand alone, without depending on other system components and without affecting the rest of the system should the one subsystem fail.

- For monitoring purposes only, it should, in addition to standing alone, be integrated with process systems. It's good engineering practice to separate execution of dissimilar functions and to integrate monitoring so that all the information is available when action is taken. It may be desirable to implement safety functions so that they can't be changed once they are in service.

Figure 9-7. Typical offshore work in progress

Typical ESD applications

Shutdown procedures are usually initiated for one (or more) of four reasons:

- Hazardous weather conditions
- Machinery failure
- Collision
- Environmental conditions (e.g., fire, or gas leak)

Shutdown logic can be tailored for specific conditions. A classic example is the operation of platforms in hazardous weather. You might want to design your ESD and process control systems to maintain unattended production in heavy weather, gradually phasing down operations as conditions worsen.

In the event of machinery failure, an expert-system capability can take into account affected peripheral operations to ensure that an individual shutdown will not create further adverse operating conditions. If that is unavoidable, the system may then select the safest method of total shutdown.

Vessel automation

As offshore exploration and production have gone into deeper water, vessels that float (in at least part of the life cycle) have taken over from fixed platforms. In many cases, platform automation is vessel automation.

Of course, there are many applications that remain strictly the province of floating systems, whether those be supply ships, drilling barges, or platforms like the Conoco Tension-Leg Well Platform (TLWP) in the U.S. Gulf of Mexico.

Vessel automation can be as complex and comprehensive as platform automation. Several semisubmersible drilling rigs in the Norwegian North Sea have integrated systems that include such applications as machinery supervision, propulsion control, power management, standby pump logics, thruster control, cargo handling, ballast and bilge control, drilling automation and winch control. The system handles more than 3,000 measuring and control points.

Main propulsion system

An automated system can monitor, control, and alarm main engine running conditions and also track fuel, lubrication and cooling systems to provide basically unattended engine room operations. Typical monitored parameters include fuel temperature and viscosity, tank levels, strainer differential pressure, lube-oil temperature and pressure, coolant temperature and pressure, shaft direction and rpm, and so on.

The system can be designed to control operations itself, or to plan operations and notify an operator. The operator should be able to observe the status of the main engines and all subsystems using graphic displays that show real-time data. If the operator does not respond to requested actions within a specified time frame, the system can take over and execute corrective action.

Diesel engine efficiency performance monitor

In addition to keeping the engine room running, an automated system can also monitor and optimize engine performance to provide energy savings and reduce downtime.

This type of system would monitor such variables as engine rpm, fuel pressure, cooling-water level and pressure, lube-oil pressure and temperature, reduction-gear-oil pressure and temperature, control-air pressure, cylinder temperature, exhaust-gas temperature, fuel injection lever position, sea-water pressure, and so on. It could:

- Analyze the engine combustion process so that it could be tuned for the most efficient operation.
- Provide long-term data for trend analysis. Utilized in an SPC approach, this information could keep engines operating within control limits, which could help prevent inefficient operation and downtime.

- Identify developing problem situations so that preventive maintenance can be performed.
- Record and store data on engine conditions and performance for future comparative purposes.

Throttle control

Automated throttle control (where the system accepts operator commands from a central console and alters engine performance accordingly) can ensure safe, reliable control of the speed and direction of the vessel's shafts. By including a speed feedback loop, it should be possible to regulate shaft speed to within 1 rpm of target speed, helping to prevent excess hunting in high seas.

An automatic shaft-overload prevention routine can prevent overloading and tripping the diesel generators. If power demand exceeds generator capacity, the control system can automatically reduce power of the propulsion motors to a level where total power demand is within generator capability. The system should be able to respond fast enough to preclude the generators tripping from overload.

Power generation and distribution

This application can start and stop each diesel generator from the central console. In addition, the automated system can monitor generator performance and handle emergency shutdown, overspeed, high coolant temperature, or over/under voltage. In vessels with more than one generator, the system can allocate resources to provide most efficient power generation and ensure that no generator exceeds its rated capacity.

In an emergency situation, the system can restrict nonessential power users and monitor power to critical systems.

Vessel motion prediction

Operating a vessel safely requires an ability to predict the vessel's response to deck loads, mooring payout and tensions, riser stresses, and the impact of wind, wave and current. Automated systems can:

- Predict the magnitude of rig motions and analyze whether or not it can safely perform a task.
- Evaluate the stability range of a rig for particular loadings.
- Monitor the rig during operations to ensure that motion does not exceed preset safety guidelines.
- Predict the behavior of the mooring system and the effect of thruster assist.
- Process weather forecasts and predict the response of the vessel.

Such systems are in use on various types of vessels, including jackups, semisubmersibles, and dynamically positioned vessels.

Load management

Load management may apply to floating platforms moored in a fixed location or vessels handling cargo, either moored or underway. A good example is the load management application on a floating platform facility. The DCS can include five process control units and more than 2500 inputs and outputs. It is comprised of three subsystems:

- The ballast and bilge control system. This is the primary means of controlling and adjusting the weight and center of gravity of the platform. The system includes the tanks, pipes, pumps, and valves that store and distribute ballast water, as well as the appropriate sensors and actuators.

  Each column of the platform has an independent ballast system that is connected tothe other columns by a continuous "ring" within the hull of the platform. Each ballast system has five tanks: Once the platform is in place, four of these tanks are evacuated and sealed off to provide flotation. The other tank remains as a ballast tank to maintain the required weight and center of gravity of the platform. This tank can take in seawater from the central ring or from the ocean, and it can also discharge water to the ring or ocean.

The ballast system is controlled by CRT consoles that include dynamic graphics, alarm summary and management, and trends. Individual tanks can also be controlled by a local control panel, and ballast valves can also be operated manually.

- Tendon tension measuring system. The platform is moored to the bottom by four cables, or "tendons," under tension. This tension is dynamically balanced against the natural buoyancy of the platform to keep the platform situated within a tight area on the ocean surface. Under wave and wind action, the platform will move in an ellipse, which must stay within preset control limits. The system continually monitors the tension on the tendons, and if they start to become overstressed, it automatically causes the ballast and bilge control system to reballast.

- Load management system. The load management system serves as the main operator interface. It furnishes the operator all the information needed to maintain safe ballasting when conditions change on the platform. Ballast tank levels, drafts, and tendon tensions are automatically input into this system. In addition, the operator can enter data about changing deck loads, and the system will calculate the required ballast transfer to obtain a certain draft and center of gravity.

  When the system was being towed to location and moored in place, the load management system operated in a "free-floating" mode to maintain an even keel under changing load conditions.

On ocean-going vessels, load management will be geared to maintaining the stability of the vessel while taking on cargo or while underway. Ballasting is a basic operation, mostly opening and closing valves and monitoring tank levels. When you deal with the dynamics associated with hanging a weight off one end of a barge and compensating for that with fluid transfer, it becomes much less simple. And when you add in wind and wave motion, the equation becomes very complex. Once the process starts to go wrong, it's not easily corrected.

One example of this type of automated load management application is a large derrick barge. Here a load management DCS with eight remote process control units monitors and controls list and trim relative to crane load, mooring-line tension and distance out, current tank levels, desired load position, vessel motion, and other factors.

In predictive ballasting mode, the system allows the operator to plan and conduct a single, smooth crane lift, or to carry out the lift in steps.

The system also functions as a data logger and alarm system for the total barge. And it acts as a "stability simulator" for operator training in disaster control.

Positioning

Positioning can refer either to station-keeping (holding a position for deep-water drilling), or tracking (laying a pipeline in the correct location). Virtually all positioning applications are automated, because positioning has to be exact. The maximum movement for a drilling vessel is 5-6% of water depth. To accommodate vessel gangways, the vessel should not move more than 20 feet in any direction.

An automated positioning system reacts not only to present conditions of the ship and the wind and waves, it projects what is going to happen 30 seconds from now, given the weather, the sea, and the ability of the ship's thrusters to respond and turn the ship. The interrelationships between all these factors are complex and critical. In a floating drilling or production system, positioning and power generation take precedence over all other functions.

The automated positioning system accepts input from environmental sensors that provide wind speed and direction, wave height, and vessel location references. It analyzes the actual position versus the desired position and generates control sequences for the thrusters. The system can automatically compensate for one or more thrusters being nonoperational.

In anchored systems, the automated system controls rotatable thrusters that aid the anchor lines in maintaining position. Such a system uses anchor-line tension and gyro-compass information about heading as input to a mathematical model that computes thruster direction and force necessary to counter the vessel's motion. The system also provides alarms for high line tension, low line tension or breakage, heading error relative to command heading, position error relative to command value, and required thrust exceeding a preset percentage of generator power.

Many positioning systems use function codes, which are predefined engineering algorithms or mathematical equations for specified analysis and control tasks. This approach provides considerable flexibility. To add an application (pipelaying, thruster mooring assist, lifting stability, station keeping), you simply add the correct predetermined function code configuration to the existing base system. The function-code DCS is built to "self adjust" to the new set of relationships.

Drilling

An important advance in drilling is steerable drilling systems. Steerable systems not only allow you to drill many wells from a single drilling platform, they also offer significant benefits in reservoir economics: A vertical well through a producing formation may have only 10 feet of casing actually exposed to the formation's pay zone. But a well drilled horizontally using steerable systems may have 10 times the exposure to the formation, with a concomitant increase in production. Some experts predict that in the future 30% or more of the world's oil and gas wells may be drilled with steerable systems.

With the advent of steerable systems, the role of automation drilling has increased dramatically, because it offers the best way to keep track of the location of the drill bit. New measurement-while-drilling (MWD) systems consist of three main components:

- Downhole sensor package. Located just above the drill bit,

MWD sensors monitor drilling parameters and even formation characteristics. Most sensor packages include directional drilling monitoring-inclination (angle), directional (azimuth), and tool-face orientation when the string is not rotating. Other sensors that may be included measure formation resistivity and gamma ray, annulus temperature and pressure, and downhole weight-on-bit and torque. The logs from these measurements permit correlation, bed boundary definition, detection of hydrocarbons and pore pressure analysis.

- A method to transmit data to the surface during drilling operations. There are two primary methods of data transmission from the downhole head assembly to the surface—pulse telemetry and continuous-wave telemetry—both of which use the drilling fluid as the medium of transmission.
- Data transmission is comparatively slow (on the order of 1-3 bits per second) when drilling fluid instead of electronics is used to transmit data. However, this is fast enough to get tool-face updates every 5-10 seconds, with weight-on-bit and torque readings every 30-60 seconds.
- Surface equipment to receive the data and put it into usable format. MWD systems generally include a pressure transducer for signal detection, analog pressure recorder, electronic signal decoding equipment, and various digital and analog readouts and plotters.

Of growing importance is software to interpret results. Today, microprocessors in MWD systems provide a good picture of the process taking place in the hole. And there are even greater opportunities for automated systems to control drilling. The next step will be the development of software to analyze drilling situations and outline prioritized changes or corrective action for every variety of downhole problem. We believe that expert systems like this will become indispensable in maximizing the return from steerable drilling systems. The benefits of having an "expert" present to interpret well logs and control drilling on a minute-by-minute basis could also make these systems useful even in simple vertical wells.

Figure 9-8. Typical offshore platform

Training and simulation

One of the real benefits of automated systems is the training and simulation opportunities they provide. If you are going to use an automated system to interface with operators on systems with a significant degree of complexity, it only makes sense to ensure that the automated system has the capability to simulate the operation for training purposes while it is also monitoring and controlling the actual operation. Using a DCS for simulation training is much more cost-effective than building a mock-up.

Simulation routines can also be helpful in managing real-life situations. For example, in a derrick barge, a ballasting simulator can also be used as a predictive ballasting system by feeding real-life environmental data. In attempting to lift unusual loads, simulating the lift first is a valuable planning aid.

Chapter 10

EXAMPLES OF OFFSHORE PLATFORM AUTOMATION

Although the offshore automation industry is still in its infancy, offshore producers have many projects completed or underway that offer examples of future directions.

We'll start with a general approach to implementation, looking first at retrofits and then at the design of new platforms. Then we'll describe five examples of automation projects completed or underway at this writing.

Retrofit projects

Retrofits are often more complex than new systems, because they must deal with an installation environment that was never designed for automated systems. Most existing platforms in the Gulf of Mexico for example, have been there for 20 years, long before today's technology became available. Even relatively simple differences in design can cause problems: For example, a retrofit platform may have no space for a central control room. Conversely, new projects can accommodate design which allows for necessary space for central control rooms from the beginning. Most of the existing control methods in the offshore industry are based on pneumatics, hydraulics, simple analog electronics, and older direct digital computer control systems. As platforms have aged, so have their systems.

This creates maintenance problems of several types. Older equipment needs frequent attention and is often harder to maintain. People often must be specially trained to carry out maintenance on old or obsolete equipment. Spare parts are hard to find, and systems cannot easily be expanded. These problems are compounded by the remoteness of the offshore environment.

Some of these operations can be made more profitable through appropriate use of new automation techniques.

But where to start? The opportunities can be overwhelming. Questions arise such as:

- Can a system be implemented one piece at a time, or must I have the entire final scheme in mind before starting?
- Where is the best place to start? Sensors? Process loops? The central control strategy?
- Where do I find the engineers who understand the system?
- How do I train operators and change over to a new system?
- How much production time will I lose doing this?
- What are the immediate monetary savings when the system is up and running?
- What are the potential long-term savings in terms of improved production time, safety, environmental concerns, and so on?
- Will automation pay for itself within a reasonable period of time?

While it's not possible to give a detailed cookbook that will cover all situations (or even most of them), here are three basic guidelines:

- Define your problems and objectives. Know what you want to achieve. Recognize that you are probably limited in this because of inadequate knowledge of the possibilities, so keep an open mind.
- Contact vendors. This is such a new field that vendors are the best overall source for finding out what can be done. Contact as many vendors as you have the time and patience to deal with.

- Look for open-ended solutions that start simple.

Much more can be said about that last point. You want to start with something that can offer immediate savings, which can then be used to fund more advanced automation. But you don't want your first choices in automation to limit your options down the road. Ask the vendor questions like:

- How does a device or system tie into higher-level systems?
- What kind of computing systems does it interface with?
- What kind of communications capabilities does it have, and what limitations?
- How much vendor support is provided for design, installation, and maintenance?
- Does the vendor have an overall philosophy on platform automation? (If not, look elsewhere.)
- What is the vendor's overall philosophy on platform automation? How does this particular device or system fit into that big picture? Do you agree?
- What is the financial strength and stability of the vendor? Will the company really be there to support you in future years? Do they have the financial stability to support continued research and development?

One logical place to start automating is at the sensor and transmitter level. More benefits can be realized by enhancing basic information about the process than in any other area. We particularly recommend that you consider "smart" and "wise" transmitters, which can usually be purchased for around $1,000 each. These transmitters provide more accurate data (which can translate directly into greater profitability), and they eliminate the need for human operator attention in the plant. "Smart" and "wise" transmitters are also self-testing, self-diagnosing, and fault-tolerant, which saves down-time and maintenance time.

The next step up is the single-loop controller and programmable logic controller (PLC). Both of these devices are proven elements that can pave the way for more complex process-control strategies. The installed cost is usually low, and the training required is usually minimal. Today, many of these units are intelligent devices, available as fault-tolerant components within a larger system.

Regardless of how you start with the above pieces, keep in mind that you want to be open to the future, in whatever guise it arrives. A PLC that costs next to nothing is worth exactly that if it does not communicate with other devices and remain compatible with future systems.

New construction

Clearly, it's much easier to incorporate today's most advanced technology when you start out correctly.

First, do your homework with the vendors. Find out what the viable options are. Weigh the costs against the benefits, being sure to include not only immediate tangible benefits, but also long-term and intangible benefits such as maintenance, liability, safety, and environmental consequences.

Then make sure that the contractor designing your platform employs not only process engineers but process *control* engineers, control-systems engineers, and communications engineers. It does no good if you understand the automation options, and the platform design engineers do not. Many engineering firms are inclined to sacrifice long-term profitability and control for lowest installed cost.

Look for a contractor with sophisticated facilities in computer-aided design, computer-aided engineering, and artificial intelligence systems. Today, designers and engineers can design an entire platform within a design computer, to give you a technologically advanced system that offers the greatest possible profitability while conforming to all applicable engineering codes and governmental regulations.

Demanning: a place to start

Unmanned offshore oil and gas production platforms are no

longer something of the future—more and more companies are considering this alternative now. This is true not just for new construction, but even in areas where benefits can be achieved from demanning an existing platform.

Many of the demanning projects that have been undertaken are those where a group of small production platforms has been demanned, and control transferred from each platform to a single central platform or a nearby on-shore operations center. These projects usually work best when initial demanning situations are on fairly unsophisticated platforms.

A typical scenario is a production platform with 3-9 wells (some of which may be water injection wells) and a single test separator. The platform would have a sea-line for outgoing oil or gas. Some of the platforms might serve as collection platforms, with an incoming sea-line or two and the single outgoing line.

The automation required on these platforms usually includes:

- A safety shutdown system (it may be possible to incorporate older pneumatic and hydraulic logic in the demanned design, or the old hardware may need to be scrapped);
- Fire and gas detection;
- Process control for the separation equipment; and
- A metering system for the various pipelines.

New distributed control systems make demanning a real option because they can handle all of the functions previously handled by several subsystems, and they are far more reliable and easy to maintain. A distributed system also makes communication with other platforms much easier to implement, and it keeps the future open for new developments.

Introduction to the case histories

As you know from earlier chapters, there are almost as many degrees and levels of automation in the offshore world as there are platforms.

Figure 10-1. A work boat servicing an offshore platform

The complexity of the remote facilities has a direct influence on the level of automation. On some platforms, automation will include extensive data gathering and process control. On others, a simple remote terminal unit (RTU) will be sufficient to open or close a valve, start or stop motors/pumps and monitor properties such as pressures, temperatures and flow.

We will now describe specific automation case histories that should give you an idea of the practical importance of automation. We'll cover:

- Retrofitting for unmanned operation in the Adriatic
- Automation on manned platforms in the North Sea

- Unmanned control of gas production and treatment in The Netherlands
- Advanced unmanned operations in the Gulf of Mexico
- One company's look at the future

Also, please note that we describe these projects as we know them in 1990. It is certain that each of these installations has, like every large integrated process facility, continued to change over time.

All of the information contained here is available in the public domain, and no information of a confidential or proprietary nature is being released or discussed.

Retrofitting for unmanned operation in the Adriatic

The first development in the Adriatic gas fields was with the installation of eight single-well platforms and a single gathering platform in the 30-foot waters. The gas was sent to an on-shore treating plant for final processing before entering the distribution network.

At first, all platforms were continuously manned. However, starting in the early seventies, they began the first attempts at platform demanning. By 1975, they had decided that all new platforms would be designed for remote control and operation, and later in the decade, they undertook a step-by-step demanning of all existing platforms.

Today, about 60 structures have been installed in the northern Adriatic Sea. All platforms connect to one of three on-shore treatment plants. All offshore facilities are now fully automated and operate unmanned, year-round. The only human attention required is a routine maintenance visit once every 48 days. The availability factor for these platforms is now greater than 98.5%.

The process scheme is essentially the same for nearly all of the platforms, which employ a single wellhead separator for each well stream. This is necessary because the reservoirs have many thin-layer production levels, and a dedicated production stream for each well allows them to maintain constant control so that each well operates within its optimal parameters.

Moreover, the single separator approach increases the platform's availability, because a problem condition in any one well can be isolated and identified immediately without compromising the other production. The separators operate at wellhead pressure because the choke valve is installed downstream of the vessel. This lowers the glycol injection rates and avoids pollution of the produced water. Produced water is discharged into the sea after treatment reduces hydrocarbon content to below 15 ppm. In some installations, produced water is sent ashore through sea-lines, which simplifies platform operations even further. Glycol is supplied to the platform through a-3 inch line from shore.

In designing automation systems for these platforms, particular attention was paid to the electrical power requirements. Electrical power via cable from shore proved expensive and unreliable due to damage caused by fishing and other activities in the area. So a series of "low energy platforms was created." On most unmanned platforms, the electrical power consumption is so low that photovoltaic cells (to 400 watts) or thermoconverter generators (to 1500 watts) can be used. Backup batteries guarantee at least six days of operation in case of a malfunction. These batteries can be recharged by an on-platform 15 KW diesel engine, which can be remotely started via the telemetry system.

All platforms and the treatment plant can be controlled remotely by redundant operator stations on shore. A radio link connects the central control desk to the offshore control systems on each platform. During manned periods, the platform can be controlled locally using the on-platform control system. Today, a modular control and supervisory system can manage 75 platforms and twelve treatment and/or storage centers. The system manages digital signals, analog signals, graphic displays, and other status points. The heart of the system is two superminicomputers operating in real-time and configured for full redundancy. These computers work in a hot-standby mode, automatically exchanging master and slave functions; while peripherals and I/O are automatically assigned to the master through automatic switching circuits. The shore-based control centers communicate to the platforms using a radio-link.

The control rooms have been organized with four operator posts. Each has two CRT monitors and an operator keyboard. The control rooms also contain color-video projectors to display plant graphics.

Other CRTs bring real-time information to the offices of reservoir managers, operation managers, and maintenance managers. In fact, special algorithms within the system help manage production, reservoir depletion, and maintenance planning. These algorithms are extremely efficient because they operate in a real-time environment with actual data.

Real-time data helps management in many ways:

- **Production management** benefits because validated production data becomes part of a single relational database on each platform. The system automatically produces periodic production reports and transmits these reports to the onshore center and to the corporate management information system.
- **Reservoir management** is also easier because of the database. Field reports are processed automatically and compared with previous models to determine the wells' real performance and total field-delivery capacities. Actual production data from each well is compared to a predetermined set of optimum well-production characteristics, which allows immediate intervention in the case of an upset. Information stored in the database even simplifies decisions regarding well maintenance.
- **Maintenance planning** is enhanced because statistical analysis can be performed on data coming in from the field to determine if the instrumentation is functioning correctly. The fault-tolerance of the field devices markedly reduces the time and expense of maintenance. The system also tracks maintenance work as part of an automated maintenance/work-order system.

Consolidating automation and control functions into a management system has greatly increased operational efficiency and decreased manpower requirements, while at the same time providing the capability for rapid intervention in the case of an emergency.

Automation on manned platforms in the North Sea

A large production facility in the North Sea is comprised of two jacketed platforms connected by a bridge—an 'A' platform for drilling and a 'B' platform for production. Each platform is comprised of a number of modules prefabricated on shore and lifted into place, and each module has a local technical room from which the systems in that module can be controlled. Although highly automated, the platforms are not unmanned. They operate without shore control, but transmit data to shore via microwave.

The process control system used on the 'B' production platform is a microprocessor-based distributed control system (DCS). The processing equipment in each module is part of a local plant loop controlled from the local technical room. Each of these local loops includes a remote termination unit (RTU) that is connected to the central control room via redundant communication links. All process control equipment is designed to continue monitoring and controlling the process even without communications from the central control console. The process control system carries out the following tasks:

- Digital and analog control
- Data acquisition, historical data storage, and event reporting
- Platform system displays (trending, alarms, etc)
- Interfacing with three other platform systems:
  - Fire and gas detection (F&G)
  - Emergency shutdown (ESD)
  - Metering.

All four systems operate separately, but share information and can be controlled from a single operator control console in the

Figure 10-2. A typical offshore installation

central control room. Operator interface units include two CRTs for process control, one for semigraphic and logging displays (reservoir management), one for metering, three for the fire and gas detection system, and one spare.

The DCS interrogates the ESD system via a common data highway and interlaces that information with information received from the RTUs in each module. The operator can also send instructions to the ESD system to initiate various levels of shutdown.

The F&G system has dedicated operator interface units for its displays, but it also communicates with the DCS via the common data highway. Shutdown instructions from the F&G system travel directly to the ESD system on the data highway.

Certain large units, such as turbine-driven compressors and pumps, have dedicated unit control panels that interface with the nearest process equipment cabinet, thereby accessing the data highway. This provides data to an operator on the plant floor and also provides control from the central control room.

A separate supervisory minicomputer also on the data highway serves as a data-handling system for metering and process data. It can provide reports on the platform and also transmit data to shore via microwave.

Central and satellite platforms in the North Sea

This facility consists of an integrated production/quarters platform connected by a bridge to a wellhead platform. The central facility also controls three unmanned satellite platforms and is linked to an on-shore terminal.

A distributed control system on the production platform provides facility-wide process control, data acquisition, event logging, and platform system displays by collecting data from RTUs on the central and satellite platforms. RTUs on the satellite platforms communicate via redundant microwave links.

The DCS also interfaces with the fire-and-gas and emergency-shutdown systems via redundant communication links. All ESD and F&G displays can be seen on the DCS' CRTs.

The separate metering system has its own dedicated operator control panel offshore and transmits data directly to on-shore computers every ten minutes. Data storage and manipulation is provided by two redundant minicomputers offshore and two redundant computers on-shore, connected by microwave.

The offshore computers are "submasters," the on-shore computers are "masters." Both offshore and on-shore, the computers operate in hot-standby mode, in which the on-line computer continually updates the standby, and the standby continually monitors the on-line machine. A failure in the on-line computer causes the standby to take control automatically.

The offshore on-line submaster continually updates the on-shore master, which is responsible for actual control of the platform most of the time. For offshore startup, or during communication failures, etc., a key-operated switch on-shore will pass control to the offshore operator.

All displays that can be seen on the offshore computer can also be seen on the on-shore computer, although the reverse is not true.

Unmanned control of gas production and treatment

The implementation of a land-based distributed control system for unmanned offshore production control includes:

- An unmanned wellhead platform for gas production with limited processing. The platform is offshore in shallow water, about 15 km from control facilities.
- The on-shore gas field.
- An on-shore gas treatment plant.

Two pipelines, a 20-inch gas pipeline and a 3-inch glycol line, connect the platform with the gas treatment plant. Because the platform is so close to shore, electrical power is provided by cable from shore. In the same cable bundle, a fiber-optic cable (chosen because fiber-optic transmission is unaffected by nearby electrical currents) provides communications between platform and plant. A radio link provides backup for communications.

All facilities are operated from the central control room at the on-shore gas treatment plant.

Due to environmental considerations, the company used a small platform with processing facilities limited to a test separator, manifold, glycol-injection system, sphere launcher, pig receiver, and utility and safety systems. Unmanning further reduced the platform size.

The DCS manages subsystems from different vendors, integrating them under control of the main system at the gas treatment plant. The DCS collects data from RTUs and PLCs at various remote locations. A typical RTU in the system has approximately:

- 135 analog inputs
- 20 analog outputs
- 355 discrete inputs
- 35 discrete outputs

The DCS controls:

- 16 wellheads and associated manifolds
- Test separator
- Glycol injection system
- Pipeline sphere launcher
- Fire water system
- Drain and vent system
- Gas oil and power generation system
- Utility and instrument air system

The DCS also transmits data regarding:

- Fire and gas detection system
- Deluge system
- Halon system

The DCS communicates with a superminicomputer, which handles database management, reporting, supervisory tasks and operation assistance.

Under normal operating conditions, all functions on the platform are operated remotely from the control room at the gas treatment facility. Well flows can be adjusted, the test separator can be monitored, and the platform can be put on temporary standby if required. Emergency shutdowns and other operations can also be initiated remotely. Display units at the platform provide a human interface for manned operation of the platform if required.

Figure 10-3. Two connected platforms

Advanced unmanned operations in the Gulf of Mexico

In 1985, a company purchased a digital distributed control system for use in an offshore production facility featuring a central platform and up to five potential remote platforms, each having approximately 800 control points. Located on the central platform, the DCS has the capability of handling that platform's production control requirements as well as controlling the remote offshore oil and gas production platforms. The system also provides the capability to control both central and remote platforms from on-shore.

The manned central platform is linked to the remote platforms and to the on-shore control center by microwave. At present, the central platform uses a PC to communicate with the land-based control center's mini-computer, also used for other functions.

One of the major considerations in this installation was alarm management, because the operating company expects this facility to continue operations during severe weather. According to their estimates, the automation system could be completely justified if the platform could be operated one day longer in the face of inclement weather. The DCS would not only have to continue operating unmanned platforms, it would have to be able to carry out the entire production shutdown sequence and the well shutin procedures.

To make this type of operation possible, fast and complete communications between platforms were imperative. And in fact, this installation uses breakthrough communications technology.

DCS systems that span several platforms use local plant loops on each platform to collect and process data. Then summary data is transmitted via communication links.

In this system, all platforms and the on-shore locations operate as part of one very large plant loop. That meant that microwave communications actually had to be imbedded within the loop, rather than acting as communication links between loops. This in turn dictated large database capability within the operator interface units, the computers connected with the system, and the microwave communication links.

To handle the microwave communication challenge, special hardware was developed to allow the integration of up to 32 plant-loop "pieces" connected by microwave into a single overall plant loop. This design now provides uniform communication and information transfer with similar operational instructions from anywhere within the system. With the introduction of packetization of digital messages in the microwave message transmission, up to sixteen exception reports can now be embedded in a single message, allowing fast transmission of large amounts of information.

Since the central platform already existed when the decision to automate was made, another important consideration was the physical size and footprint of the DCS. Space was at a premium.

Another advanced technology, high-density termination units, made it possible to connect 16 to 20 wires in a space that would normally only accommodate 4 to 6. Ultimately, the DCS was installed in less space than had originally been occupied by the conventional controls.

One final concern was how to install a new system with operators who were not familiar with the software languages resident in the on-shore computer system or with the function-block code of the DCS. Fortunately, a DCS that allows operators to input logic in a ladder-logic format was used. Operators easily understand this system, and it allows them to operate, troubleshoot, and maintain the platforms more efficiently.

One company's look at the future

This is a research project aimed at determining the possibility of, and defining the specifications for a central unmanned production platform. The platform is to be remotely controlled and visited only once annually for maintenance and servicing.

The intermediate objectives are the functional design of pilot subsystems and the testing of new equipment that may be installed on the platforms.

The project emphasizes automation as a means of long-term profitability. It was started in early 1986, and the first platform is scheduled to be put on-stream in 1992-1993. Specifications will be issued late 1990, early 1991. The annual research budget amounts to about $50 million. The goals of this massive research project are:

- Improved oil field development through the control of operating conditions (reliability and control).
- The development of marginal fields and keeping these fields on-stream using conventional recovery methods.
- Production optimization (reduced downtime, optimal operation).

- Better job engineering (control jobs substituted for routine operational jobs).
- Improving safety conditions through a reduction of offshore personnel

This project clearly reflects a commitment to innovate. Some of the new technology they have expected from the project includes:

- More reliable sensors for offshore operations.
- More reliable data gathering capability, using redundant data processing software.
- On-line expert systems for use in maintenance and operations.
- New processes developed for production systems.
- New high-tech rotating machinery.
- Robot-aided monitoring systems.
- Systematic reliability studies on components and systems

Summary

The description of these facilities and systems has been necessarily concise, and many aspects have had to be omitted or only briefly mentioned. We hope, however, that we have included enough information to give an overall view of how DCS systems operate in the real world.

Clearly, automation is a growing trend in the offshore industry. Automation is the only means of improving productivity and building a secure economic footing for future operations. And it is the only means of avoiding loss of life and skyrocketing liability as exploration and production move into more environmentally challenging areas.

We believe that distributed control systems have an expanding role to play in offshore automation. A DCS makes the most of new technology as it comes about. It is expandable for virtually any size operation, without the limitations inherent in PCs, minis, and mainframes. It can "pull in" whatever components are needed for a particular job, from simple sensors to large mainframe computers. And because it is distributed, wherever there is work to do a DCS is the most flexible approach to real-time control applications, now and in the future.

Figure 10-4. A distributed control system

INDEX

V

W

NOTES

NOTES

NOTES

NOTES

NOTES

NOTES

NOTES

NOTES